REASONING, COMMUNICATION AND CONNECTIONS IN MATHEMATICS

Yearbook 2012

Association of Mathematics Educators

REASONING, COMMUNICATION AND CONNECTIONS IN MATHEMATICS

Yearbook 2012
Association of Mathematics Educators

editors

Berinderjeet Kaur
Toh Tin Lam

National Institute of Education, Singapore

Published by

World Scientific Publishing Co. Pte. Ltd.
5 Toh Tuck Link, Singapore 596224
USA office: 27 Warren Street, Suite 401-402, Hackensack, NJ 07601
UK office: 57 Shelton Street, Covent Garden, London WC2H 9HE

British Library Cataloguing-in-Publication Data
A catalogue record for this book is available from the British Library.

REASONING, COMMUNICATION AND CONNECTIONS IN MATHEMATICS
Yearbook 2012, Association of Mathematics Educators

ISBN-13 978-981-4405-41-6
ISBN-10 981-4405-41-8

Printed in Singapore by B & Jo Enterprise Pte Ltd

Contents

Chapter 1

Reasoning, Communication and Connections in Mathematics: An Introduction

Berinderjeet KAUR TOH Tin Lam

This introductory chapter provides an overview of the chapters in the book. The chapters are organised according to three broad themes that are central to reasoning, communication and connections. The themes are mathematical tasks, classroom discourse and connections within and beyond mathematics. It ends with some concluding thoughts that readers may want to be cognizant of while reading the book and also using it for reference and further work.

1 Introduction

This yearbook of the Association of Mathematics Educators (AME) in Singapore focuses on Reasoning, Communication and Connections in Mathematics. Like two of our past yearbooks, Mathematical Problem Solving (Kaur, Yeap, & Kapur, 2009) and Mathematical Applications and Modelling (Kaur & Dindyal, 2010), the theme of this book is also shaped by the framework of the school mathematics curriculum in Singapore, shown in Figure 1. The primary goal of school mathematics in Singapore is mathematical problem solving and amongst the processes specified explicitly for nurturing problem solvers are reasoning, communication and connections. In elaborating the framework, for both the primary and secondary students, the Ministry of Education (MOE) (2006a, 2006b), syllabus documents clarify that:

Mathematical reasoning refers to the ability to analyse mathematical situations and construct logical arguments. It is a habit of mind that can be developed through the application of mathematics in different contexts. Communication refers to the ability to use mathematical language to express mathematical ideas and arguments precisely, concisely and logically. It helps students develop their own understanding of mathematics and sharpen their mathematical thinking. Connections refer to the ability to see and make linkages among mathematical ideas, between mathematics and other subjects, and between mathematics and everyday life. This helps students make sense of what they learn in mathematics. (Ministry of Education, 2006a, p.14; 2006b, p.5)

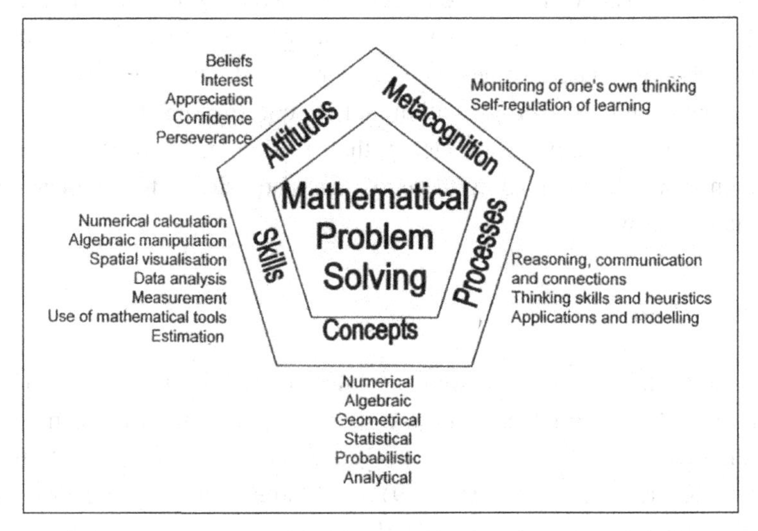

Figure 1. Framework of the Singapore school mathematics curriculum

The fourteen peer-reviewed chapters in this yearbook address various aspects of reasoning, communication and connections. All the chapters arose out of the keynote lectures and workshops conducted during the Mathematics Teachers Conference 2011 which was jointly organised by the Association of Mathematics Educators and the Mathematics and Mathematics Education Academic Group at the National Institute of Education in Singapore. The authors of the chapters were asked to focus on evidence-based practices that school teachers can experiment in their lessons to bring about meaningful learning outcomes. In the following sections we briefly outline how the chapters contribute towards reasoning, communication and connections respectively.

2 Mathematical Tasks

Central to all mathematics lessons are mathematical tasks. A mathematical task is defined as a set of problems or a single complex problem that focuses students' attention on a particular mathematical idea (Stein, Grover, & Henningsen, 1996). From the TIMSS Video Study (NCES, 2003), in which Australia, Czech Republic, Hong Kong, Japan, Netherlands, Switzerland, and the United States participated, it was found that students spent over 80% of their time in mathematics class working on mathematical tasks. According to Doyle (1988), "the work students do, defined in large measure by the tasks teachers assign, determines how they think about a curricular domain and come to understand its meaning" (p. 167). Hence different kinds of tasks lead to different types of instruction, which subsequently lead to different opportunities for students learning (Doyle, 1988).

Mathematical tasks with high cognitive demand often require students to make explicit their thinking. These tasks are necessary for the advancement of reasoning, communication and connections during lessons. In chapter 2, Rahim, Hogan, and Chan report research on the status of epistemic framing of mathematical tasks in secondary three mathematics classes in Singapore. The findings reported in the chapter indicate that teachers provided students with relatively substantial opportunities to acquire procedural and metacognitive knowledge but not conceptual and epistemic knowledge. As the intellectual quality of knowledge work depends substantially, although not exclusively, on the nature and quality of tasks that teachers' provide their students to work with, it is apparent that teachers need knowledge of such tasks themselves.

Boston and Smith (2009) report that research has consistently indicated that teachers selection of instructional tasks is largely based on lists of skills and concepts they need to cover. Textbooks are often the main source of such tasks (Doyle, 1983; Kaur, 2010). As noted by Valverde, Bianchi, Wolfe, Schimdt, and Houang (2002) textbooks are often good guides for content but lack emphasis on processes such as problem solving or communication. In this book there are several chapters that provide teachers with insights on how to modify textbook type of mathematical tasks into ones that incorporate reasoning and facilitate communication in the lesson. In chapter 3, Thompson presents a variety of specific and general strategies that may be used to modify typical textbook exercises to incorporate reasoning and communication into the primary mathematics classroom. In chapter 4, Kaur introduces primary teachers to four "What" strategies that may be used to create tasks from textbook questions or student work so as to advance reasoning and communication in the primary school mathematics classrooms. The strategies are: What number makes sense?, What's wrong?, What if?, and What's the question if you know the answer? In the chapter how one primary 1 teacher enacted the strategy "What's the question if you know the answer" is also presented to demonstrate that even primary one students can participate in lessons that call for reasoning and communication provided they are given the opportunity to do so. In chapter 5, Thompson draws on a framework from textbook research and shows how typical algebra and statistics textbook type of exercises may be modified to engage secondary school students in reasoning and justification. The framework has six components of proof-related reasoning, viz-a-viz finding counterexamples, investigating conjectures, making conjectures, developing arguments, evaluating arguments, and correcting mistakes in reasoning.

Lew and Jang, show in chapter 6 that project-based LOGO programming activities may be used to improve reasoning, communication and problem solving skills of high ability Year 6 students. These activities appeared to provide students with opportunities to activate and promote reasoning strategies such as analogy, generalisation, progressive and critical thinking and debugging based on visualisation and empirical inference. In chapter 7 Toh illustrates with appropriate examples, from the Advanced Level Mathematics curriculum for Years 11 and 12, how reasoning, communication and connections may be infused into the teaching of mathematics for advanced learners of mathematics. Toh demonstrates that for mathematical reasoning, elementary mathematical proofs and derivation of results by first principles are good opportunities to transcend procedural emphasis of traditional teaching to higher level mathematical reasoning. He also illustrates that connections can be achieved by several means such as facilitating students to connect across different mathematical ideas, across other disciplines and to daily life.

Lastly, in chapter 8, Lowrie uses assessment tasks to make the case that visual and spatial reasoning plays an important role in communicating mathematical ideas. Furthermore he emphasises that the increasing reliance on graphics in today's society require students to acquire different spatial-reasoning skills so that they consider all the elements of a task, including specific features of a graphic and the surrounding text, when solving mathematical tasks.

3 Classroom Discourse

Mathematical tasks mediate classroom discourse between teachers and students, and between students and students. As stated in the Professional Standards for Teaching Mathematics (National Council of Teachers of Mathematics, 1991), "the discourse of a classroom – the ways of representing, thinking, talking, agreeing, disagreeing – is critical to what students learn about mathematics as a domain of human inquiry with characteristic ways of knowing" (p.34). Teacher questions are a key element of classroom discourse. Such questions provide students interactive opportunities for learning. Hogan, Rahim, Chan, Kwek and Towndrow in chapter 9 report research on the kind of teacher questions that teachers focus on in mathematics lessons and the relationships between different kinds of teacher questions. Their research is based on survey data from students in secondary three mathematics classes in Singapore. The findings of the research show that there is a relatively high prevalence of performative questions arising from the IRE [teacher initiate (I), student respond (R), and teacher evaluate (E)] talk structure in the mathematics classes. This suggests that mathematics classrooms in

Singapore provide limited opportunity for students to engage in rich classroom conversations.

Analogical reasoning draws on pre-existing knowledge to make relational and structural similarity in objects and construct mathematical knowledge. Lee in Chapter 10 shows that analogical reasoning can facilitate rich discourse, both teacher initiated and student initiated, between the teacher and students. Student errors and misconceptions are often outcomes of classroom discourse where students fail to clarify their thinking or teachers adopt "recipe like" procedures that are not conceptually sound. Pang and Dindyal in chapter 10 studied students' reasoning errors in proofs by mathematical induction. The errors made by the Year 12 students mainly arose from their inability to grasp that in the mathematical statement $P(n)$ n must be an integer and that the set of numbers generalised by n can be ordered. Also, many students regard the induction step solely as proving the $P(k + 1)$ statement and not proving that "if $P(k)$ is true then $P(k + 1)$ is true". It is apparent that the errors were a result of perception that mathematical induction is a procedure and not a coherent deductive system of logical steps of a proof.

4 Connections Within and Beyond Mathematics

Among the aims of mathematics education in Singapore schools are:

- Acquire the necessary mathematical concepts and skills for everyday life, and for continuous learning in mathematics and related disciplines;
- Recognise and use connections among mathematical ideas, and between mathematics and other disciplines.

(Ministry of Education, 2006a, p. 2; 2006b, p. 11).

It is apparent from the above that connections within and beyond mathematics are an essential part of the school curriculum. Teachers must provide students with opportunities to experience connections in the mathematics they learn. This is possible through links between conceptual and procedural knowledge, connections among mathematical topics and equivalent representations of the same concept (Coxford, 1995). Similarly, teachers must also provide students with opportunities to experience connections between mathematics and other disciplines of the school curriculum and daily life needs.

Leong in chapter 12 provides a portrait of mathematics classroom instructional practice in which mathematics is presented as skills and concepts that are closely connected to each other. Here, students are encouraged to see school mathematics as situated and connected to a wider context of mathematical knowledge. Leong discusses ways in which this alternative portrait of mathematics instruction can be enacted, drawing upon examples that include findings from a research project. In school education, numeracy is a fundamental component of learning, discourse and critique across all areas of the curriculum. In chapter 13, Kissane discusses the nature of numeracy and some recent work in the Australian primary and secondary curriculum that involves numeracy. According to Kissane, at the primary levels, situations that demand mathematical thinking arise in all aspects of the curriculum and thus provide opportunities for numeracy to be developed and at the secondary levels, mathematics is important for learning in other subjects.

Kemp in chapter 14, asserts that mathematics teachers often find that resources for teaching and learning, such as standard textbooks, do not connect well to the everyday worlds of students in their classes. In the chapter, Kemp shows that there are many opportunities in the classroom to make connections between school mathematics and the everyday world using the example of health. Lastly, Aslaksen in chapter 15 shows how astronomy and culture may be connected to mathematics learning in classrooms of students from primary schools to university.

4 Some Concluding Thoughts

The performance of students from Singapore in the Trends in International Mathematics and Science Study (TIMSS) 1995, 1999, 2003 and 2007 and the Programme for International Student Assessment (PISA) 2009 has been outstanding. This affirms that the basics of mathematical content knowledge are sound. However, the research reported in this book on the epistemic framing of mathematical tasks and classroom talk in secondary three mathematics classes in Singapore show that the quality of knowledge work to facilitate mathematical practices in the classrooms were lacking. As, so aptly, stated by Ball (2003):

> Mathematical practices involve more than what is normally thought of as mathematical knowledge. This area focusses on the mathematical know-how, beyond content knowledge, that constitutes expertise in learning and using mathematics. The term "practices" refers to the specific things that successful mathematics learners and user do. Justifying claims, using symbolic notation effectively, defining terms precisely, and making generalisations are examples of mathematical practices. Another example of mathematical practices is the way in which skilled mathematics users are able to model a situation to make it easier to understand and to solve problems related to it. Those skilled individuals might use algebraic notation cleverly to simplify a complex set of relationships, or they might recognise that a geometric representation makes a problem almost transparent, whereas the algebraic formulation, although correct, obscures it. (p. xviii)

Although, it is apparent from the Singapore school mathematics framework (shown in Figure 1) that the intended curricula places emphasis on processes such as reasoning, communication and connections, thinking skills and heuristics, and applications and modelling for the development of "practices" that Ball (2003) refers to, there appears to be lack of emphasis on these in lessons. To improve the quality of knowledge work to facilitate mathematical practices, teachers need to engage their students in rich mathematical tasks and classroom discourse.

The several chapters in the yearbook provide readers with ideas on how to improve the quality of knowledge work in mathematics classrooms through reasoning, communication and connections. We suggest you read the chapters and contextualise the ideas within the larger context of teaching and learning mathematics. We also like you to note that reasoning, communication and connections are not isolated sub-components of Processes in the framework, rather they work in tandem with other components of the framework. For example, the use of appropriate questioning techniques, modification of textbook questions to engage students in reasoning and communication, use of visual and spatial reasoning for solving mathematical problems also help students regulate their learning and engage in reflection i.e. metacognition. Similarly, when teachers make connections across different mathematical ideas and disciplines of the school curriculum students get an opportunity to see the application and relevance of the mathematics they learn. This may help to nurture positive attitudes towards the learning of mathematics for some students.

Lastly, we like to offer a word of caution for readers who are often overwhelmed by yet more ideas in the book in view of their already content-heavy curriculum. We urge the readers to read the chapters carefully and try some of the ideas in their classrooms and convince themselves that these ideas offer a means of infusing reasoning, communication and connections in their lessons and engage students in meaningful mathematical practices.

References

Ball, D.L. (2003). *Mathematical proficiency for all students: Towards a strategic research and development program in mathematics education*. Santa Monica: Rand.

Boston, M.D. & Smith, M.S. (2009). Transforming secondary mathematics teaching: Increasing the cognitive demands of instructional tasks used in teachers' classrooms. *Journal for Research in Mathematics Education*, *40*(2), 119–156.

Coxford, A.F. (1995). The case for connections. In P. A. House & A.F. Coxford (Eds.), *Connecting mathematics across the curriculum*. Reston, VA: National Council of Teachers of Mathematics.

Doyle, W. (1983). Academic work. *Review of Educational Research*, *53*, 159–199.

Doyle, W. (1988). Work in mathematics classes: The context of students' thinking during instruction. *Educational Psychologist, 23*, 167–180.

Kaur, B. (2010). A study of mathematical tasks from three classrooms in Singapore schools. In Y. Shimizu, B. Kaur & D. Clarke (Eds.), *Mathematical Tasks in Classrooms around the World* (pp. 15–33). Sense Publishers.

Kaur, B. & Dindyal, J. (2010). *Mathematical applications and modelling*. Singapore: World Scientific.

Kaur, B., Yeap, B.H., & Kapur, M. (2009). *Mathematical problem solving*. Singapore: World Scientific.

Ministry of Education, Singapore (2006a). *Mathematics syllabuses – Primary*. Singapore: Author.

Ministry of Education, Singapore (2006b). *Mathematics syllabuses – Lower secondary*. Singapore: Author.

National Council of Teachers of Mathematics (1991). *Professional standards for teaching mathematics*. Reston, VA: Author.

NCES (National Center for Educational Statistics). (2003). *Teaching mathematics in seven countries: Results from the TIMSS video study*. Washington, DC: U.S. Department of Education.

Stein, M. K., Grover, B., & Henningsen, M. (1996). Building student capacity for mathematical thinking and reasoning: An analysis of mathematical tasks used in reform classrooms. *American Educational Research Journal, 33*, 455–488.

Valverde, G. A., Bianchi, L. J., Wolfe, R. G., Schmidt, W. H., & Houang, R. T. (2002). *According to the book: Using TIMSS to investigate the translation of policy into practice through the world of textbooks*. Dordrecht, The Netherlands: Kluwer Academic Publishers.

Chapter 2

The Epistemic Framing of Mathematical Tasks in Secondary Three Mathematics Lessons in Singapore

Ridzuan Abdul RAHIM   David HOGAN   Melvin CHAN

This chapter describes two aspects of knowledge work that are important in considering instructional tasks in classrooms: the epistemic focus of knowledge work and epistemic (knowledge) practices. We focus on four types of knowledge: epistemic, procedural, conceptual and metacognitive, and six types of knowledge practices: knowledge communication (use of syntax), knowledge communication (presentation), representation, generation, deliberation and validation. Using data from a survey of over 1000 Secondary 3 students in 30 schools drawn from a representative random stratified sample of secondary schools in Singapore, we report on: 1) the kind of mathematical knowledge(s) that teachers focus on in mathematics classrooms; 2) the kind of disciplinary knowledge practices teachers ask students to engage in; and 3) the relationships between different kinds of knowledge and practices. The students reported prevalence of metacognitive and procedural knowledge as well as knowledge communication (use of syntax) in the instructional tasks that teachers structured for them. However, they also indicated that teachers do not generally provide students with a rich and integrated array of opportunities to engage in knowledge practices at a level and in a sequence that emphasizes the disciplinary structure of mathematics work. We suggest that a clearer awareness of the epistemic nature of knowledge claims and instructional tasks that attends to the structure and logic of knowledge practices in mathematics could render instructional tasks more "authentic" — and *doing* mathematics more transparent, "visible" and accessible.

1 Introduction

After the choice of learning goals, the choice of instruction and assessment tasks are the most important instructional decisions that teachers make. In principle, these choices should always precede the choice of instructional methods: good instruction practice requires us to treat instructional methods as a means to an end, not ends in themselves, and to ensure proper alignment of methods with goals and tasks. The ends are given by learning goals and the instruction and assessment tasks that operationalize them. Normatively speaking then, tasks are pivotal to the overall character of instruction conceived of as a dynamic, interdependent and hierarchically organized system rather than as a collection or aggregation of relatively discreet, independent instructional practices. This is a very different approach to the one that appears to principally frame the discourse of teaching in Singapore — a tendency to *frame* questions about teaching in terms of questions about the curriculum, the scheme of work, activities, methods and classroom environment rather than in terms of *instruction* as a systematic hierarchical ordering of classroom practices.

Still, it's one thing to develop challenging learning goals, and entirely another to design assessment and instructional tasks that give expression to these goals and facilitate their realization. Indeed, we take the design of assessment and instructional tasks, the former especially, along with the quality of classroom talk and the appropriateness of the instructional methods, to be one of the three critical determinants of the intellectual quality of teaching and learning in the classroom. In a way, we might see this *task* as an effort, ironically enough, to "bring knowledge back in" (Young, 2008) to lesson planning and the instructional process in a coherent and rigorous way. We think the benefits of a renewed focus on tasks, and on the knowledge work embedded in tasks, will be quite substantial.

The classic study of the nature and role of instructional tasks is Walter Doyle's influential 1983 paper on "academic work" in the classroom. Doyle claimed that academic tasks "form the basic treatment unit in classrooms" and were, therefore "the primary determinant of how the curriculum is experienced by students" and the principal feature of classroom life that mediates between teacher instructional behavior and student learning. Tasks "influence learners by directing their attention to particular aspects of content and by specifying ways of processing information. These effects are clearly apparent in the contrast between semantic and non-semantic processing... the processing of information for meaning versus the processing of information for surface features" (p. 161). Although Doyle recognizes that other classroom events and processes influence student learning, Doyle insists that academic tasks serve as the proximal causes of student learning: "Tasks influence learners by directing their attention to particular aspects of content and by specifying ways of processing information" (pp. 160-161).

From Doyle's perspective then, the tasks with which students become engaged determined not only *what* content they learn but also *how* they can engage in the academic work of the classroom and therefore learn how to think about, develop, make sense of and apply the content knowledge they encounter in the task. Consequently, whereas some tasks engage students at a surface level, others engage students at a deeper cognitive level by demanding interpretation, flexibility, the shepherding of resources, and the construction of meaning. Of course, as Stein and her colleagues have argued over many years (Stein, Grover, & Henningsen, 1996; Henningsen & Stein, 1997), the intellectual quality of the tasks that teachers design ("task set up") are not always, even rarely, matched by the intellectual quality of the knowledge work that students complete ("task implementation"). But despite evidence of weak task fidelity, there is now a substantial body of research that identifies the task activities as the key determinant of the quality of the intellectual work that students engage in. For example, in his recent meta-meta-analytic study of student achievement, Professor John Hattie (2009, p. 22) write that "It is what teachers get students to do in the class that emerged as the strongest component of the accomplished teachers repertoire, rather than what the teacher, specifically does." In effect, in learning, what teachers do matters, but what students do matters even more. Similarly, Newmann, Bryk and Nagaoka (2001, p. 31) in reporting on the large Annenberg Project in Chicago, indicate that the key to securing high quality learning outcomes depends not so much on the selection of instructional methods but on the construction of high quality instructional tasks. "We wish to emphasize," they write, that "no particular teaching practice or strategy assures that students will undertake work that makes high-quality intellectual demands on them... *Our key point is that it is the intellectual demands embedded in classroom tasks, not the mere occurrence of a particular teaching strategy or technique, that influence the degree of student engagement and learning."* Elsewhere, in research reported sometime earlier, Newmann and Associates (1996) argue that the kind of tasks that have the greatest impact on student learning are those characterized by disciplinary depth and connection to "authentic" intellectual standards outside the school.

In this chapter, we want to focus on what we term the *epistemic framing* of instructional tasks in Secondary 3 Mathematics. In particular, we want to investigate the epistemic *focus* of knowledge work in the classroom, and we want to investigate the nature of the epistemic (knowledge) *practices* that teachers ask students to engage in through the tasks they set for students. The former refers to the kind of knowledge that teachers focus on — factual, conceptual, procedural, epistemic, metacognitive — while the latter refers to the kinds of domain specific (or disciplinary) knowledge practices that teachers require students to engage in from generating and representing knowledge claims through to deliberating and validating knowledge claims. In a sense this distinction parallels a related distinction between knowledge as a noun and cognition as a verb implicit in the well-known taxonomy of knowledge and cognition developed by Anderson and Krathwohl (2001, pp. 4-5). In this chapter, however, we are interested in marrying epistemic verbs to epistemic nouns. In later papers we will report on the *cognitive* and the *normative* framing of instructional tasks, where the former focuses on the nature of the cognitive demands that mathematics tasks require, and the latter on the nature of the task values (mastery, competitive) that characterize specific tasks and the kinds of epistemic virtues that teachers seek to encourage in asking students to engage in particular kinds of knowledge work.

Broadly speaking, our interest in the epistemic framing of instructional tasks reflects a broader theoretical interest in understanding the intellectual quality of knowledge work in Singaporean classrooms. Singaporean students obviously do very well in international assessments (TIMSS, PISA) although the range of tasks associated with these assessments are relatively narrow. In particular, it is far from clear to us that either assessment or instructional tasks clearly reflect the intrinsic disciplinarity of domain-specific knowledge and knowledge practices associated with specific disciplines, or that they go much beyond testing student memorization and understanding of mathematical knowledge at the expense of their understanding of mathematical processes and knowledge practices. Conceptually, we might think of a discipline as a public and relatively coherent (if contested) *body of historically situated knowledge* (facts, concepts, propositions, analytical relationships and networks, arguments, models, theorems) *and* institutionalized *knowledge practices*, procedures, skills, epistemic norms, dispositions, identities and attachments for investigating, generating, representing, communicating, deliberating, justifying, evaluating and/or applying knowledge claims *focused variously on* the explanation, prediction, interpretation, representation, expression, understanding or control of some aspect of human experience (the natural world, the social world, ourselves) (Hogan, 2007). School subjects are not disciplines as such, although they are influenced by certain features of disciplinary knowledge and practices, particularly in high school. Rather, school subjects are latent or compound constructions shaped by a range of curriculum principles including disciplinarity, developmental considerations, bureaucratic concerns and political commitments, and teacher *habitus*. Still, notwithstanding the range of demands that need to be accommodated in the construction of school subjects, the degree of disciplinarity of school subject knowledge is a leading indicator of the intellectual quality of knowledge work in the classroom (Hogan, et al., 2011). In particular, disciplinarity provides a useful framework for reconciling knowledge as a noun and knowledge as a verb (as a process or a practice).

In this chapter we draw on this model of disciplinarity to inform our analysis of the epistemic framing of instructional tasks in Secondary 3 mathematics. In particular, we ask what kind of mathematical *knowledge*(s) do teachers focus on in class, what kind of disciplinary knowledge *practices* do teachers ask students to engage in, and what is the *relationship* between the two? Or, to put it in slightly different terms, what epistemic nouns do teachers focus on, what epistemic verbs do teachers insist on, and how well do the verbs and nouns work together? In effect, we focus on two *epistemic* features of instructional tasks in Singaporean mathematics classes: the epistemic focus of the content knowledge in instructional tasks and the knowledge (or epistemic) practices associated with generating, representing, deliberating, communicating, justifying and applying knowledge claims. It is important to recognize, however, that these two dimensions of instructional tasks are *only* two aspects of instructional tasks, since there are a number of others as well, including, in particular, clarity about learning goals and the cognitive processes that the task is intended to engage students in, that together constitute instructional tasks but we are not able to canvass in this chapter.

The data on which this chapter draws is based on the Core 2 Research Program at NIE. Specifically, it principally draws on survey data from Secondary 3 students in all the Secondary 3 classes in a stratified random representative sample of schools across the system.

2 Epistemic Framing 1: Knowledge Focus

In developing our taxonomy of knowledge, we have relied substantially on the well-known taxonomy of cognition and knowledge developed by Anderson and Krathwohl (2001) and other — factual knowledge, conceptual knowledge, procedural knowledge and metacognitive knowledge. Briefly, **factual knowledge** is *propositional* knowledge (dates, events, facts, names, equations, definitions, algorithms, and etc.). Propositional knowledge is either true or false. **Procedural knowledge** is knowledge that focuses on *how* an epistemic agent undertakes and completes a task specific to a discipline, subject or area of study. It can refer to quite general procedural issues — methods of inquiry, particular methodologies, genres of work — or, more narrowly, to task-specific scripts, strategies, algorithms, heuristics involved in solving a problem or generating knowledge claims. Hiebert and Lefevre (1986), for example, suggest that "procedural knowledge consists of knowledge of (a) the formal symbol system of mathematics [syntactic knowledge] and (b) rules, algorithms or procedures used to solve mathematical tasks." On this account, solving a particular kind of problem, strategies, algorithms, heuristics, norms/genres for generating knowledge claims constitute procedural knowledge in mathematics. **Conceptual knowledge** focuses on the *meaning* of concepts and *relationships* between concepts e.g., tree and forest, light and gravity, revolution and nationalism, sincerity and authenticity, square and triangle, etc. etc. Conceptual knowledge thus includes semantic knowledge — the meaning of words — but also, critically, relationships, patterns, networks and connections between ideas rather than propositional knowledge (discrete bits of information or factual knowledge). **Metacognitive knowledge** is a form of reflexive knowledge of the self that focuses on how one learns and the development of what Maurice Galton (2007, ch. 5) terms "metacognitive wisdom." It thus focuses on monitoring one's learning and developing knowledge of effective strategies and heuristics that "works" for the student. Finally, **epistemic knowledge** is knowledge of domain-specific criteria and standards that allows individuals ("epistemic agents") to distinguish "knowledge" from mere information, opinion or belief by appeal to domain specific criteria — truth, reliability, validity, coherence, authenticity, clarity, reasonableness, principled, disinterested, goodness, or beauty — that establishes the public authority of a knowledge claim. Epistemic knowledge then above all is concerned with the identification and justification of the standards of justified belief or knowledge.

We report the descriptive statistics of the knowledge focus items and scales we have employed in Table 1. The indicators for conceptual and procedural knowledge are reasonably good while those for metacognitive and epistemic knowledge are reasonable enough, reflecting the *post hoc* nature of our analysis of instructional tasks in classrooms: when we designed the project in 2009, we did not quite anticipate all the theoretical opportunities that the data would open up for us. For this reason we do not have a scale for factual knowledge. This is especially unfortunate, given our sense of the importance of factual knowledge in Singaporean mathematical classes — for example, knowledge of definitions and algorithms. Be that as it may, metacognitive knowledge (m=3.59), surprisingly, had the highest mean score followed by procedural knowledge (m=3.48). Epistemic knowledge (m=3.36) and conceptual knowledge (m=3.29) and came in third and fourth, respectively.

Overall, the absolute levels of the mean scores are low to moderate on a five-point Likert scale, indicating some room for a stronger emphasis on conceptual and epistemic forms of mathematical knowledge especially. The *relative* strength of metacognitive knowledge is especially welcome, given the importance of meta-cognition in developing the key capability of leaning to learn and by Schoenfeld's (1992) description of the actions of an expert problem solver who employs metacognitive self-regulation to solve problems, as opposed to novice problem solvers who tend to continue using the same method to solve a problem even when it does not lead to success. The relative strength of procedural knowledge was not a surprise in the Singaporean context, although we anticipated that its mean value would be higher than what it is here, given the purported emphasis on procedural knowledge in East Asian pedagogy. The relative weakness of conceptual knowledge was an unwelcome surprise, given that research on cognition and mathematics strongly suggests that conceptual understanding of mathematical notions and relationships is fundamental to *doing* mathematics well and to enhancing the disciplinarity of knowledge work in mathematics classrooms. We were not especially surprised by the mean score for epistemic knowledge, but it is a relatively thin measure of epistemic knowledge and, like conceptual knowledge, leaves substantial room for improvement. Overall, teachers' emphasis on metacognitive knowledge and procedural knowledge suggests that instructional tasks are focused on getting students to learn how to solve problems or complete the tasks and with less emphasis on getting a deep understanding of the concepts and by what standards the knowledge can be justified.

Table 1
Descriptive statistics of knowledge focus in secondary 3 Mathematics

Knowledge Focus	Alpha	Mean	S.D.
1. Procedural Knowledge	.889	3.48	.756
The teacher emphasizes understanding of formulae by asking us to apply them to different kinds of problems.		3.52	.912
The teacher teaches us how to identify the best method to solve problems.		3.56	.901
The teacher emphasizes using alternative ways to solve a problem.		3.43	.894
The teacher provides opportunities for us to adapt MATHS procedures to new problems.		3.40	.915
The teacher teaches problem-solving strategies (e.g. working backward, drawing a diagram).		3.49	.921
2. Conceptual Knowledge	.891	3.25	.764
The teacher asks us to analyze problems or information.		3.38	.908
The teacher asks us to explore a mathematical concept in depth.		3.20	.945
The teacher asks us to think deeply to understand an important topic very well.		3.34	.941
The teacher asks us to apply what we have learned to practical issues or real life situations.		3.15	.991
The teacher asks us to solve and present detailed solutions to difficult problems.		3.33	.926
The teacher asks us to perform tasks that require you to link ideas from different subjects.		3.09	.992
3. Metacognitive Knowledge	.727	3.59	.797
The teacher says we must try to understand how we learn.		3.56	1.019
The teacher teaches us to be aware of the strategies we use in solving problems.		3.63	.945
The teacher asks us to check if our answers make sense?		3.57	1.008
4. Epistemic Knowledge	.767	3.36	.862
The teacher says our arguments must be correct.		3.07	1.119
The teacher says our reasoning must be logical.		3.60	1.013
The teacher says our reasoning and expressions must be consistent with mathematical concepts.		3.41	1.002

(N=1438)

To test the reliability and factor structure of the scales, we ran a series of exploratory and confirmatory factor analyses of the four scales. Both forms of factor analysis strongly supported the four scale structure reported in Table 1; in particular, the goodness-of-fit statistics for a hierarchical CFA were well above (or below) the appropriate thresholds. We provide a detailed report of the CFA statistics elsewhere (Hogan, et al., 2011). Table 2 reports the correlations between the latent knowledge constructs based on the CFA. The strength of the correlation between procedural and conceptual knowledge (.777) is especially significant, given the controversy within mathematics as to whether procedural knowledge nurtures conceptual knowledge, vice versa, or some other possible relationship. Putnam, Lampert and Peterson (1990, pp. 84-85), for example, insist that at least in mathematics "conceptual competence" or conceptual knowledge depends on an understanding of procedural knowledge: "in some cases, procedural knowledge must form the basis for conceptual understanding." On the other hand, for Rittle-Johnson & Alibali (1999), teaching children the concept behind mathematical equivalence problems, rather than a procedure for solving them, was most effective at promoting flexible problem-solving skill and conceptual understanding. Children who have a better understanding of place value are more likely to successfully use the borrowing procedure for multi-digit subtraction, for example (Cauley, 1988; Hiebert &Wearne, 1996)[1]. Yet a third possibility is that procedural and conceptual knowledge iteratively inform each other (Hiebert & Lefevre, 1986) — that is, that procedural knowledge aids conceptual knowledge and vice versa. A simple bivariate correlation doesn't resolve this conundrum, but it does indicate that there is a strong relationship between the two. We will come back to this issue later.

[1]Hiebert & Wearne (1996) cited in Rittle-Johnson & Alibali (1999)

Table 2

Correlation matrix of latent constructs of knowledge focus in secondary 3 Mathematics

	Procedural Knowledge	Conceptual Knowledge	Metacognitive Knowledge	Epistemic Knowledge
Procedural Knowledge	1			
Conceptual Knowledge	.777**	1		
Metacognitive Knowledge	.554**	.522**	1	
Epistemic Knowledge	.456**	.426**	.745**	1

*= *Statistically significant at .01 level*

While useful for indicating the degree of association between the various measures of mathematical knowledge, correlation coefficients do not tell us anything either about the causal direction of the association or the strength of the relationship controlling for the influence of confounding variables. Structural equation modeling (SEM) is a sophisticated statistical technique that allows us to address both of these concerns by establishing how well hypothesized pathways in the model in fact account for empirical relationships in the data (technically, fitting the theoretical or structural variance/covariance matrix to the measurement variance/covariance matrix). We report the results of our best fitting SEM model in Figure 1. The model fit statistics show a reasonable although not exceptional fit. Intriguingly, this particular model identifies epistemic knowledge as the form of knowledge that supports (predicts) conceptual knowledge. However, surprisingly, there is no direct path from epistemic knowledge to conceptual knowledge or procedural knowledge. A strong direct path from epistemic knowledge to metacognitive knowledge is obtained (regression coefficient = .74). Importantly from a theoretical perspective, the gatekeeper for this model is metacognitive knowledge. Metacognitive knowledge has a direct path (.55) to procedural knowledge and both an indirect (.379) and direct path (.14) to conceptual knowledge. Table 3 shows the indirect effects of epistemic knowledge on conceptual knowledge (.280) and procedural knowledge (.407); and from metacognitive knowledge to conceptual knowledge (.379).

All of these indirect effects are moderate and significant. In addition, two important observations can be made. First, the direct path between epistemic knowledge and conceptual and procedural knowledge is non-significant while the indirect path is significant. In mediation analysis, metacognitive knowledge is said to fully mediate (MacKinnon, 2008) the effect from epistemic knowledge to procedural knowledge and conceptual knowledge. This underscores the importance of metacognitive knowledge on both procedural and conceptual knowledge with respect to epistemic. In other words, *the effect of epistemic knowledge on conceptual knowledge and procedural knowledge would be negligible without metacognitive knowledge*. Second, going from metacognitive knowledge to conceptual knowledge, there are both significant direct and indirect (i.e., through procedural knowledge) paths. Procedural knowledge is said to moderate (MacKinnon, 2008) the effect of metacognitive knowledge on conceptual knowledge. So here we see the importance of procedural knowledge in bringing about conceptual knowledge.

Table 3

Indirect and direct effects for knowledge focus model (Figure 1)

Unstandardized Parameter Estimates	Indirect	Direct	Total Effects
Epistemic Knowledge → Conceptual Knowledge	.280	0	.280
Metacognitive Knowledge → Conceptual Knowledge	.379	.14	.519
Epistemic Knowledge → Procedural Knowledge	.407	0	.407

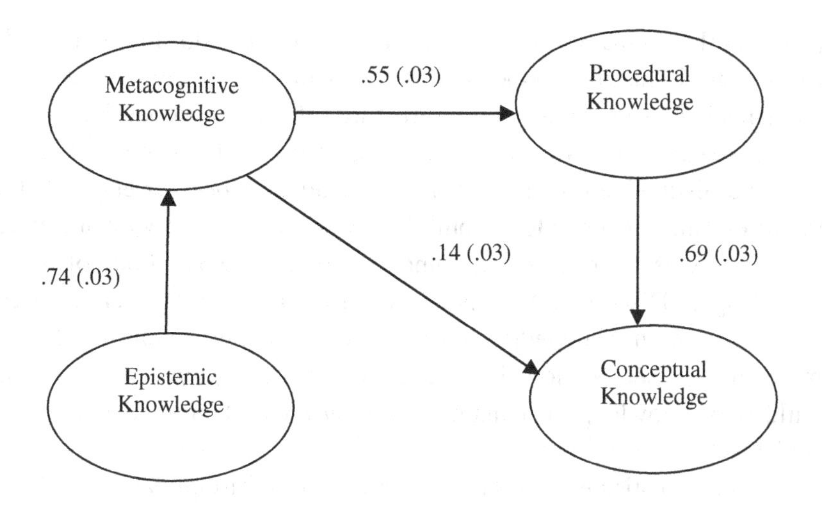

Figure 1. SEM model of knowledge focus: Secondary 3 Mathematics

Goodness-of-fit statistics	
Chi-Square / df / p-value	3.967 /2 /.1376
CFI / TLI	.999/.996
RMSEA (90% CI)	.029 (.000-.071)
SRMR	.012

Note: This SEM model was constructed using composite unstandardized latent constructs.

3 Epistemic Framing 2: Domain-Specific Knowledge Practices

In selecting tasks, teachers need to choose not just the epistemic focus of the tasks they want students to engage in, but what kind of domain specific or disciplinary knowledge practices they want students to engage in as well. Sometimes they do this with a reasonably developed model of disciplinary knowledge work in mind, sometimes not. Sometimes they pay little or no attention to knowledge practices at all, but instead focus simply on the acquisition of content and skill. When they do so, they are unlikely to engage students in that repertoire of knowledge practices that helps define the disciplinarity of the field. In other words, from a disciplinarity perspective, it is important not only to focus on mathematical knowledge as a structure of propositional knowledge *claims* — but also on the nature of the epistemic or knowledge *practices* (and the distinctive cognitive and mathematical *processes* embedded in them) essential to knowledge building both intra-subjectively and inter-subjectively as a set of personal understandings and as a body of public knowledge. Disciplinarity then involves marrying epistemic verbs (participating in knowledge practices) to epistemic nouns (bodies of more or less settled domain-specific knowledge claims) in ways that build new knowledge (in mathematics *writ* large) or in the minds of students.

In mathematics education, the importance of disciplinary knowledge practices has been highlighted, for example, by two leading US mathematics education researcher, Professor Jo Boaler and Professor Deborah Ball. First, Professor Boaler (Boaler, 2002a, p. 2; Boaler, 2002b, pp. 7-8):

> Researchers of mathematics education have, for many years, focused upon the knowledge students develop in classrooms, and the ways such knowledge is influenced by a number of different variables. Recently, situated theories of learning have led to the recognition that the practices of classrooms — the repeated actions in which students and teachers engage as they learn — are important, not only because they are vehicles for students' knowledge development, but because they come to constitute the knowledge that is produced (Cobb, 1998). *Thus the field has moved to greater recognition of the intricate relationship between knowledge and practice and the need to study the practices of classrooms in order to understand students' mathematical capability in different situations.* [Italics added]

So too Professor Ball (Ball, 2003):

> Mathematical practices involve more than what is normally thought of as mathematical knowledge. This area focuses on the mathematical know-how, beyond content knowledge, that constitutes expertise in learning and using mathematics. *The term "practices" refers to the specific things that successful mathematics learners and users do. Justifying claims, using symbolic notation efficiently, defining terms precisely, and making generalizations are examples of mathematical practices.* [Italics added] Another example of mathematical practices is the way in which skilled mathematics users are able to model a situation to make it easier to understand and to solve problems related to it. Those skilled individuals might use algebraic notation cleverly to simplify a complex set of relationships, or they might recognize that a geometric representation makes a problem almost transparent, whereas the algebraic formulation, although correct, obscures it. (p. xviii)

More generally, scholars influenced by research on the nature of expertise and by social-cultural learning theory have particularly emphasized the idea of learning as a *cognitive apprenticeship* via *situated cognition* in *communities of practice* and the importance of participating in domain-specific knowledge *practices* as well as learning bodies of (more or less contentious) disciplinary knowledge (Brown, Collins, & Newman, 1989; Resnick, Saljo, Pontecorvo, & Burge, 1997; Lave,1988; Lave & Wenger, 1991; Tharp & Gallimore, 1998; Engle & Conant, 2002; Engle & Faux, 2006; Ford & Forman, 2006; McConachie & Petrosky, 2010, pp. 21-22). McConachie and Petrosky (2010, p.10), for example, argue that while content knowledge matters, so do knowledge practices or knowledge tasks: the kinds of knowledge tasks that students are asked to engage in, and the kind of interactions they have in class, matter a great deal to the development of what they term "disciplinary literacy." "All our work proceeds from these principles [among others] and the notion that disciplinary knowledge always coexists with habits of thinking... particular to each discipline," they write. "*Teaching and learning in the disciplines, then, involves students in doing the work of the discipline.*" They are far from alone in thinking so. At a general level, as Sawyer in the recent *Handbook of the Learning Sciences* (2006, pp. 1-2) notes, the learning sciences have demonstrated that "students cannot learn deeper conceptual understanding simply from teachers instructing them better. Students can only learn this by actively *participating* in their own learning" (emphasis added) although I would add they that they do so best when they are engaged in authentic tasks in properly structured roles. John Hattie, similarly, argues on the back of his meta-meta-analysis of student achievement that "it is… what learners *do* that matters." Effective teaching makes "students active in the learning process — through actions by teachers and others — until the students reach the stage where they become their own teachers, they can seek out optimal ways to learn new material and ideas, they can seek resources to help them in this learning, and when they can set appropriate and more challenging goals." Indeed, Hattie goes on, "students need to be involved in determining success criteria, setting higher expectations, and being open to experiences relating to differing ways of knowing and problem solving. This then leads to their development of beliefs and reputations as a learner, and engaging in self-assessing, self-evaluating, self-monitoring, self-learning, and in learning the surface, deeper, and conceptual domains of worthwhile domains." (2009, p. 37).

While Hattie's argument is framed in the pedagogical language of cognitive engagement and learning rather than the epistemic language of participation in knowledge practices and knowledge building, the point is well taken. Indeed, underlying the research findings that Sawyer and Hattie summarize is an awareness that older models of learning of disciplinary knowledge in which knowledge or conceptual development are framed as processes of knowledge "acquisition," to use Anna Sfard's (1998) phase, have been superseded by newer constructivist metaphors of learning that focus on situated cognition and "participation" in communities of practice. "The set of new key words that, along with the noun 'practice,' prominently features the terms 'discourse' and 'communication' suggests that the learner should be viewed as a person interested in participation in certain kinds of activities rather than in accumulating private possessions," she writes. "To put it differently, learning a subject is now conceived of as a process of becoming a member of a certain community. This entails, above all, the ability to communicate in the language of this community and act according to its particular norms (Sfard, 1998, pp. 5-6).

The data reported below focus on domain specific knowledge practices for Secondary 3 Mathematics. Importantly, these scales are *not* intended to capture all forms of knowledge work that might occur in Mathematics. Rather, they report what kind of knowledge practices teachers ask students to engage in mapped against a normative model of key disciplinary knowledge practices. Table 4 reports the rank order of the mean scores and standard deviations by scale in Secondary 3 Mathematics. In general, the mean scores are relatively low: only mathematical syntax approaches a reasonably good score (3.66 on a 5 point scale). Knowledge communication (use of syntax) measures the extent to which instructional tasks require the use of syntactical knowledge of mathematics in communicating a mathematical idea. Mathematical communication requires the use of the correct syntax in order to understand a task, manipulate ideas within the task and to convey the solution or the output to the task. The next four scales had similar mean scores: Knowledge Representation (3.12), Knowledge Generation (3.10), Knowledge Deliberation (3.14) and Knowledge Validation (3.16). Tatsuoka et al. (Tatsuoka, Corter, & Tatsuoka, 2004) suggested that "Singapore students obtained top performance on TIMSS mainly by showing excellence in reading and computational skills." By reading they meant "understanding verbally posed questions, translating/formulating equations and expressions to solve a problem, and quantitative and logical reading." If this was true then knowledge representation should have a larger mean score than reported. However, upon further reflection of the items in the knowledge representation scale, the lower scores can be explained by the emphasis on the use of multiple representations and creating new representations (this item itself had a low mean score of 2.93, see Table 4). Critically, the use of multiple representations is emphasized by NCTM (2000) — students need a variety of representations to support their understanding. One can therefore infer that while students in Singapore are probably conversant with the practice of developing one representation appropriate for solving a problem, they have had much less exposure to using multiple representations to enhance their understanding.

Table 4
Means and standard deviations, disciplinary knowledge practices, secondary 3 Mathematics, 2010

Knowledge Focus			
	Alpha	Mean	S.D.
1. Knowledge Communication (Use of Syntax)	.769	3.66	.806
To write correct mathematical sentences		3.57	.980
To present a solution to a problem in a logical step-by-step manner		3.81	.976
To use the correct symbols at the proper places (e.g., equal signs, letters to represent the unknown)		3.43	.969
2. Knowledge Representation	.913	3.12	.756
To represent a real world problem as a mathematical problem		3.00	.965
To represent an idea in different ways (e.g., using graphs, tables or algebraic symbols)		3.20	.909
To use different representations (for e.g., symbols, tables, diagrams, graphs) to highlight differences or similarities between mathematical ideas		3.16	.931
To create your own way of representing (e.g., make your own diagram or table) a mathematical idea		2.93	.977
To use a mathematical representation of an idea you have learned in a new context or situation		3.12	.924
To change from one way of representing an idea to another so that you understand the idea better		3.22	.910
To compare different types of representations that can be used in solving a problem		3.17	.910
3. Knowledge Generation	.910	3.10	.710
To work on problems that require you to use ideas you have learnt from other mathematics topics		3.34	.914
To contribute new ideas when the teacher is teaching a new topic		3.05	.959
To contribute your ideas when working in a group		3.26	.984
To work out your own way of solving a problem		3.22	.937

To modify ideas you have learnt from solving a previous problem in order to solve a new problem		3.20	.898
To use your understanding from other subjects in order to understand a problem		3.12	.938
To use your understanding from other subjects in order to understand a new mathematical idea		3.06	.945
To use your experiences from outside the classroom in order to understand a mathematics problem you are trying to solve		2.99	.978
To use your experiences from outside the classroom in order to understand a new mathematical idea that the teacher is teaching		2.94	.970
To solve a mathematical problem without using methods your teacher has taught you.		2.79	1.046
4. Knowledge Deliberation	.832	3.14	.840
To make a guess to a solution and then discuss the reasons why the guess might be correct		3.11	.964
To make a guess to a solution and then check if your guess is correct		3.12	.986
To discuss as a class mathematical ideas or suggestions someone has made		3.18	.965
5. Knowledge Communication(Presentation)	.692	2.87	.791
To come to the whiteboard to solve a problem		3.41	1.049
To do a group presentation of what your group has discussed		2.79	1.076
To do an oral presentation or explanation of your solution to a problem		2.83	1.016
To write a journal on the mathematics that you have learned		2.43	1.249
6. Knowledge Validation	.902	3.16	.846
To comment on the correctness of the mathematical argument someone has made		3.12	.962
To correct a mathematical argument someone has made		3.18	.976
To say what is wrong with a mathematical argument someone has made		3.16	.969
To say why someone's guess to a problem might be correct or incorrect		3.17	.943

Source: Panel 2 Secondary 3 Mathematics Student Survey, Sept/Oct 2010. N=1913

As we did with epistemic focus, we also ran exploratory and confirmatory factor analyses of the scales to establish the reliability of the various scales. As we report elsewhere (Hogan, et al., 2011) the CFA generated strong goodness-of-fit statistics for the model with a clear higher order factor, strong factor loadings, and solid but not excessive latent construct correlations.

The lower mean scores for knowledge generation, deliberation and validation suggest that the instructional tasks given to students largely do not require them to participate in such disciplinary practices. It implies that instructional tasks are either fairly routine that it only requires students to use procedural knowledge and to use the correct syntax of mathematics to work out the task and communicate their findings; or that even more complex tasks are reduced in their task cognitive demands to enable students to do them. The goal of doing such tasks therefore is not for students to participate in knowledge practices to generate knowledge new to themselves but to simply be able to perform or what (Kapur & Bielaczyc, 2011) called unproductive success, i.e. "an illusion of performance without learning."

Finally, the knowledge communication (presentation) scale had the lowest mean score of 2.87. Admittedly, students coming to the whiteboard had a reasonably high mean score of 3.41. However the other three items in the scale had much lower mean scores ranging between 2.43 and 2.83 — and these had to do with either doing oral presentation or written (possibly non-mathematical) text presentation. Thus the instructional tasks given to students do not appear to provide the opportunity for students to communicate in substantive ways to share their ideas and clarify their understanding (NCTM, 2000). As (Cobb, Wood, & Yackel, 1993) noted, students do not talk about mathematics naturally and so they need to learn how to do so.

In sum: although it's possible that the relative strength of tasks focusing on mathematical syntax might help explain the success of Singapore students in international assessments in Mathematics in recent years, the overall level of opportunity for students to engage in rich disciplinary knowledge practices, particularly beyond knowledge tasks focused on learning mathematical syntax, is relatively limited. Unfortunately, this limitation is somewhat obscured by the effectiveness of conventional instructional practices in developing student competencies in the kinds of mathematical skills and understandings measured by international assessments (TIMSS, PISA). While these skills and understandings are important, in the context of 21st century knowledge economies and societies, they are arguably insufficient, and need to be augmented by a stronger disciplinary understandings and capabilities. Expanding access to a broader range of high quality disciplinary practices is not to render mathematics more arcane, abstract or academic, but to render it more accessible, accountable, *disciplined*, "authentic," transparent and "visible." For Newmann and Associates (1996, ch.1), for example, authentic knowledge work is knowledge work that "stands for intellectual accomplishments that are worthwhile, significant, and meaningful, such as those undertaken by successful adults: scientists, musicians, business entrepreneurs, politicians, crafts people, attorneys, novelists, physicians and so on." He continues: "With children we are concerned with a more restricted conception of achievement, one that can be accomplished in schools. For students, we define authentic achievement through three criteria critical to significant intellectual accomplishments: construction of knowledge, disciplined inquiry, and the value of achievement beyond the school." What our model of disciplinarity offers is an even more differentiated model of knowledge construction, disciplined inquiry and public value than offered by Newmann by focusing on quite specific kinds of knowledge practices associated with knowledge construction, disciplined inquiry and public value in mathematical knowledge work. Similarly, conceiving mathematical work in this way enhances what John Hattie (2009) terms "visible learning" (p. 25) that he proclaims *the* fundamental instructional principle.

For these reasons then our expectation (although we cannot demonstrate it at the moment) is that a mathematics curriculum that is epistemically self-conscious about the nature of mathematical knowledge and knowledge practices and that is serious about offering rich opportunities to students to engage in a full range of authentic disciplinary knowledge practices that afford the exercise of a wide array of mathematical processes will be at once more engaging, effective and relevant to 21st century institutional preoccupations with "knowledge building" and not just knowledge transmission, especially for students who otherwise might find mathematics impossibly abstract, inaccessible and boring. In a later study we plan to test this hypothesis using a longitudinal experimental design: if confirmed, it would go some way to reducing the considerable tail in mathematics achievement that plagues educational systems East and West and demonstrate, once again, the folly of assuming that equity comes at the expense of high achievement.

Efforts to shore up the presence of mathematical knowledge practices in the classroom will be supported by the very substantial correlations between the various knowledge practices reported in Table 5. (Note that these correlations are based on the degree of association between latent constructs developed in the CFA analysis rather than on the summative scales). Plainly, the correlations reported in Table 5 are generally moderately high to high, suggesting that teachers believe that these practices, in relationship to one another, are not competitive but complementary, and that, therefore, they do not have to choose between knowledge practices but can choose any or all of them without having to worry about pedagogical opportunity costs. From our perspective, this is a well-grounded belief, and ought to be strongly encouraged in in-service and pre-service programs, although it is important that teachers also understand the optimal sequence of pathways in order to maximize the disciplinarity of the knowledge practices that they engage in. But to grasp this set of understandings, we need to draw upon more complex multivariate statistical techniques.

Correlational analysis, as we indicated earlier in our discussion of Table 2, while useful as a first take on the strength of the relationship between two or more variables, does not by itself identify the direction of the association or its strength after controlling for confounding variables. We again used structural equation modelling techniques to model the structure and strength of the pathways between the various knowledge practices (Figure 2). Overall, the goodness-of-fit statistics is quite strong, indicating the structural model shadows and accounts for the measurement model exceptionally well. But as pleasing as the overall goodness-of-fit statistics are, what is especially interesting about the model are the specific pathways from the key gatekeeper practice (mathematical syntax) through to the two key outcome measures (knowledge communication (presentation) and knowledge validation) via knowledge representation. Although there are two other pathways, one indirectly via knowledge generation (.10) on to knowledge communication (presentation) (.39), and one direct from knowledge communication (syntax) to knowledge validation (.21), these two pathways are significantly weaker than the multiple ones that go through knowledge representation and far less able to expose and engage students in the full suite of disciplinary practices that our model of mathematics disciplinarity allows.

The relatively strong focus of Singapore's teachers on mathematical syntax indicated in Table 4 is supported by the Ministry of Education's mathematics curriculum framework and by our structural equation modelling of the relationship between the various knowledge tasks reported in Figure 2. The pathways and path coefficients reported in these figures strongly indicate that Singapore's Secondary 3 Mathematics teachers believe that the mastery of *mathematical syntax* is essential to *do* and *learn* mathematics by learning to engage in knowledge practices of the mathematics community more broadly.

From a theoretical perspective, this is surely correct. A critical aspect of learning mathematics is learning its language. If a student understands the language or syntax of mathematics, they are able to understand and communicate mathematical syntax. Indeed, we hypothesize that knowledge communication through the use of the correct mathematical syntax forms the basis of a mathematical knowledge practice. In addition, student facility with the use of mathematical syntax in Singapore might well help explain why Singaporean students have done as well as they have in international mathematics assessments. As Tatsuoka et al. (2004), as we reported earlier, suggests, "Singapore students obtained top performance on TIMSS mainly by showing excellence in reading and computational skills." By reading they meant "understanding verbally posed questions, translating/formulating equations and expressions to solve a problem, and quantitative and logical reading."

We can put this another way. From a communities of practice perspective, learning mathematics syntax is akin to learning the rules of the community a novice epistemic agent wishes (or is required) to join. In the mathematics classroom, the community of learners first learn the rules of mathematics grammar in order to participate in more complex mathematical practices. At this point the learners would only be admitted to the periphery of the practice. The goal of learning from a disciplinarity perspective is to participate in the full repertoire of mathematical practices and to build knowledge — conceptual, procedural, epistemic and metacognitive. This argument can also be rendered using a different (although related) metaphor of learning: from the perspective of cognitive apprenticeship theories of learning, unless students learn how to participate in mathematical practices that are closer to authentic disciplinary practices beyond learning how to manipulate syntax — representation, generation, deliberation, validation, communication — they remain novice cognitive apprentices in mathematics and are not able to graduate into the full community of mathematics practice.

Figure 2 also indicates that the teachers' assume that the principal pathway to the conceptual world of more complex forms of mathematical practice principally (although not exclusively) goes through *knowledge representation*, not direct participation in the other knowledge practices included in the model. And again, we think this reflects a well-supported research understanding of *disciplinary* mathematical practice. Knowledge of how mathematical knowledge is represented, through symbols, graphs, tables, pictorial forms, etc., facilitates an understanding of how knowledge can then be generated, deliberated, presented and validated. Indeed, Putnam et al. (1990), drawing on the work of Kaput (1985; 1987a; 1987b), especially, in their review of research in mathematics education, state that "virtually all of mathematics concerns the representation of ideas, structures, or information in ways that permit powerful problem solving and manipulation of information" (p. 68). For Putnam, as for Kaput, representation is understanding: "much of mathematics involves the representation of one mathematical structure by another and determining what is preserved and what is lost between the two structures" (p. 69)[2]. However, while the coefficient for the syntax-representation pathway is quite strong (.500), indicating that while teachers believe it is the most efficacious pathway from knowledge of syntax to participation in the other knowledge practices, the low mean score for knowledge representation (m=3.11) indicates that relative to teaching knowledge of syntax (m=3.66), Singapore teachers do not make as much an effort as they might to developing multiple mathematical representations of a mathematical concept or problem in moving from one representation to another. This is troubling, given the conviction of many mathematics education researchers, including the National Council of Teachers of Mathematics (NCTM, 1991), that multiple representations are a key gateway to conceptual understanding.

[2]See also Goldin and Schteingold (2001)

Table 5

Correlation matrix, latent constructs, disciplinary knowledge practices, secondary 3 Mathematics, 2010

Knowledge Practices	Communication (Syntax)	Representation	Generation	Deliberation	Communication (Presentation)	Validation
Communication (Syntax)	1					
Representation	.539**	1				
Generation	.519**	.791**	1			
Deliberation	.530**	.716**	.763**	1		
Communication (Presentation)	.397**	.563**	.617**	.599**	1	
Validation	.519**	.616**	.659**	.687**	.512**	1

Note: **= Statistically significant at .01 level

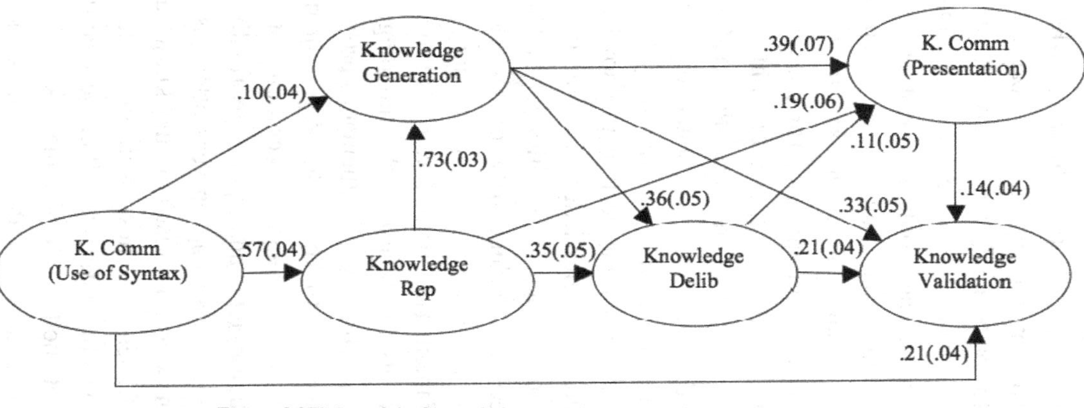

*Figure 2.*SEM model of knowledge practices: Secondary 3 Mathematics

Note: Values on the left represent unstandardized estimates significant at $p<.01$. Values on the right are standard errors.

Goodness-of-fit statistics		Scale	Mean	SD
(N=1166)		K. Com (Syntax)	3.66	.806
Chi-Square / *df* / *p-value*	3.392 / 3 / .3350	K. Representation	3.12	.756
CFI / TLI	.997 / .988	K. Generation	3.10	.710
RMSEA (90% CI)/ SRMR	.011 (.000-.052) / .006	K. Deliberation	3.14	.840
		K. Com (Presentation)	2.87	.791
		K. Validation	3.16	.846

Note: This SEM model was constructed using composite unstandardized latent construct values rather than measurement values.

Figure 2 also identifies three separate pathways from knowledge representation (the key gatekeeper to conceptual knowledge work) to knowledge validation (the key *epistemic* practice). What essentially differentiates these pathways is the range and depth of disciplinary-based knowledge practices that they give students the opportunity to participate in: some limit the opportunity to participate, others give students extended opportunity to work through the logic of disciplinary knowledge work from encoding mathematical problems in mathematical syntax to validating knowledge claims. One pathway, for example, moves directly from knowledge communication (syntax) directly to knowledge validation (.21) without any opportunity to participate in any of the other knowledge practices that together constitute a rich model of knowledge practices in mathematics. By any measure, this pathway reflects a much diminished — indeed minimalist — understanding of the disciplinarity of mathematics. A second pathway, also relatively low in disciplinarity, proceeds from knowledge communication to knowledge generation (.10) and from there to knowledge communication (presentation) (.39) and from there to knowledge validation (.14). There are also other pathways that selectively include between knowledge generation, deliberation, knowledge communication (presentation) and knowledge validation, but not all of them. The one pathway that provides the maximum opportunity for students to engage in the full range of knowledge practices central to the disciplinarity of mathematics proceeds from knowledge communication (syntax) to knowledge representation (.57) to knowledge generation (.73) to knowledge deliberation (.36) to knowledge communication (presentation) (.11) and from there, finally, to knowledge validation (.14). While it doesn't happen very often (as we can see when we check the mean scores for each of the knowledge practices and the regression coefficients to how well one practice predicts another), it is the *only* pathway that integrates all of the specified knowledge practices into a coherent model of disciplinary knowledge building. An instructional regime that had a deeper understanding and commitment to *disciplinarity* — to disciplinary knowledge work — would have far stronger coefficients as well as mean values. While the mean score for knowledge communication (syntax) is relatively high, the mean scores for all other practices are relatively low. In addition, while the coefficients from knowledge syntax to knowledge representation and from knowledge representation to knowledge generation are relatively high (.57 and .73 respectively), all the coefficients that follow are relatively small; indicating that the prior practice doesn't predict the secondary practice very well at all. Indeed, the weak regression coefficients between knowledge communication (presentation) and knowledge validation (.14), knowledge deliberation and knowledge validation (.21) and from knowledge deliberation to knowledge communication (presentation) (.11) all indicate that there is little follow through along these pathways. There are moderate pathways from knowledge representation to knowledge deliberation (.35) and from knowledge generation to knowledge communication (presentation) (.39), but beyond this, the low mean scores as well as the weakness of the coefficients suggests that teachers do not engage in these practices very often, or if they do, rarely proceed beyond knowledge representation and knowledge generation. Indeed, the weak size of the coefficients emanating from them suggests that teachers tend to focus on one of three terminal practices: knowledge deliberation, knowledge communication, or knowledge validation, but not all three together, let alone in the appropriate sequence. This is confirmed, for example, by our estimate of the covariance between knowledge communication (presentation) and knowledge validation (.022). In short, the SEM model suggests that teachers view these knowledge practices as quite separate kinds of knowledge work rather than articulated elements of an integrated disciplinary sequence. Moreover, the mean score for knowledge communication (syntax) (3.66) relative to the mean scores for the other scales indicates that teachers tend to emphasize the acquisition of the fundamental tools of mathematical knowledge building, as they should, but they do so at the expense of appropriate levels of attention to those additional knowledge practices and their interrelationships necessary for students to fully participate in mathematical communities of practice. In Anna Sfard's terms, mathematical teachers in Singapore tend to view mathematics learning as a process of "acquisition" rather than "participation" (1998). This is not to say that teachers ought to replace an emphasis on knowledge acquisition with a focus on participation in knowledge practices. Rather, as Sfard insists, teachers need to be guided by both perspectives. We shall return to this issue again later.

We also tested an alternative maximal model that ran the pathway from knowledge communication (syntax) to knowledge representation (.57) to knowledge generation (.73) to knowledge deliberation (.36) to knowledge validation (.21) and from there, finally, to knowledge communication presentation (.18) but the goodness-of-fit statistics (chi square=5.971, df=4, p=2014, CFI/TLI=999/997, RMSEA (90% CI)/SRMR = 021 (.00-.052)/.07) are not as strong as the previous model. It is also not as strong theoretically from a disciplinary perspective: whereas knowledge communication (presentation) in the former model appropriately functions as a mechanism to validate knowledge claims, in the later model knowledge validation occurs prior to knowledge communication. While the numbers tell us teachers do this, it's not optimal from a disciplinary perspective. Consequently, we strongly recommend the former pathway as the pathway most likely to maximize the opportunity of students to participate in the full range of knowledge practices that fulfil the promise of expansive models of mathematics disciplinarity. Our principal concern, from the perspective of mathematics education, is that there are four or five pathways that teachers enact in Singaporean classrooms that do not provide expansive opportunities to students to participate in the full range of knowledge practices that mathematics makes available from a disciplinary perspective and does not, therefore, fully exploit the opportunities for mathematics knowledge building.

Two final points. First, while both knowledge generation and knowledge representation are critical practices mediating between knowledge communication (syntax) and knowledge validation via knowledge deliberation, from a disciplinary perspective it is important to get the sequence right. Both have pathways into knowledge deliberation, but the theoretically stronger pathway goes from knowledge representation to knowledge generation and from there to knowledge deliberation rather than from knowledge representation to knowledge deliberation directly. This underscores the importance of knowledge generation in doing knowledge deliberation tasks.

Second, Table 6 reports the single direct and selective indirect effect sizes of the pathways from knowledge communication (syntax) to knowledge validation. Critically, the pathway from knowledge communication (syntax) to knowledge validation that optimizes participation in disciplinary knowledge practices discussed above, has one of the smallest effect sizes (.002) of all the pathways. Of course, this is in part a function of the mathematics for calculating the effect size: the more fractional coefficients you have the lower the product (.57*.73*.36*.11*.14=.002). But the low effect size is also due to the very small coefficients of the last three pathways. If the last three pathways had had the same coefficient as the second (.73), the total effect size would have been a very respectable .161, the highest of all the indirect pathways. Indeed, going forward, we would hope that implementation of an instructional improvement program (including PD) with a strong disciplinary focus would go some way in improving not just the mean scores but the coefficients as well since there is not much point in increasing the mean scores of the various knowledge practices if the articulation between them is not also improved.

Table 6

Direct and selected indirect effects for knowledge practice model

Direct/Indirect Effects	Unstandardized Parameter Estimates (S.E.)
1. K. Com (Syntax) → K. Validation	.21 (.04)**
2. K. Com (Syntax) →K. Gen→K.Com (Presentation)→K. Validation	.034 (.03)**
3. K. Com (Syntax) →K. Gen→K. Delib→K.Com (Presentation)→K. Validation	.001 (.000)^{ns}
4. K. Com (Syntax) →K. Gen→K.Com (Presentation)→K. Validation	.005 (.003)*
5. K. Com (Syntax) →K. Rep→K. Delib→K.Com (Presentation)→K. Validation	.003 (.002)*
6. K. Com (Syntax) →K. Rep→K. Delib→K. Validation	.043 (.010)**
7. K. Com (Syntax) →K. Rep→K. Gen→K. Validation	.136 (.022)**
8. K. Com (Syntax) →K. Rep→K. Gen→K Delib→ K.Com (Presentation)→K. Validation	.002 (.001)^{ns}
Total for K. Com (Syntax)→K. Validation	.513

Note: *=Significant at .05 level ** = Significant at .01 level

Overall, the results of our descriptive analysis and the structural equation models point to the broad conclusion that although teachers provide relatively high levels of opportunity to understand and use mathematical syntax, teachers also engage in a limited repertoire of critical domain-specific knowledge practices that reflect disciplinary forms of mathematic knowledge work. In short, taking into account both the mean scores and the pathway regression coefficients, while teachers give learners access to the community of mathematics practice by providing them with relatively strong background in mathematics syntax, the quite low mean scores for knowledge practices beyond syntax-related practices suggests that students tend to remain marooned at the periphery of the broader mathematics community of practice. This is unfortunate, and hardly what anybody would want.

We can state this conclusion slightly differently in these terms. Consider dividing up these knowledge practices into three sets of practices: knowledge communication (syntax) and representation; knowledge generation and/or deliberation; and finally knowledge communication (presentation) or knowledge validation. We can say that much of the work in Singapore classrooms is concentrated on the first set of practices — knowledge communication (syntax) and knowledge representation. There is some indication of participation in the second and third set of practices, but a complete repertoire of practices is not obvious from the student reported data. The second type of practices, knowledge generation and knowledge deliberation is where the heavy lifting of mathematics takes place — the processes of thinking out possible ways of interpreting the representations to solve the problem or testing out hypotheses with what-ifs and then testing them out. The third set of practices knowledge communication (presentation) and knowledge validation brings us closer to the practices of the discipline. The work of mathematicians often seen in journals involves the third set of practices. However, it must be noted that the heavy lifting work of generating ideas to solve the problems and making hypothesis and testing them are often not seen. In school mathematics we often observe a short-circuiting of the process of generation and deliberation through teaching of algorithmic procedures or steps in solving known problems. The end result is that knowledge practices tend to remain limited to the syntax and representation levels.

4 Tying the Epistemic Knot: Structural Equation Models of Knowledge Focus and Knowledge Practices

We earlier suggested that the disciplinarity framework offers a way of reconciling two key aspects of domain-specific knowledge: the epistemic focus or status of the knowledge in question and the nature of the epistemic or knowledge practices associated with the generation, representation, communication, deliberation and validation of knowledge claims. We now want to test this hypothesis formally by examining the relationship between epistemic focus and knowledge practices. In doing so we assumed that it is simply not sensible, from a theoretical perspective, to isolate knowledge as a body of propositional knowledge from the knowledge practices that generate, represent, communicate, deliberate and validate those knowledge claims, let alone give in to the temptation that constructivists have been prone to do, of collapsing knowledge (a noun) into a process (a verb). But before we do so it would be well to note Sfard's argument that a balanced and sensible theory of learning disciplinary knowledge will need to recognize that learning needs to be understood in two ways: as "acquiring" a publicly validated body of knowledge claims, and as well as a process of "participation" in a community of practice. But knowledge cannot be collapsed into "knowing," "doing" or "participation." Indeed, as she puts it, "one metaphor is not enough." (Sfard, 1998, p.16).

Table 7 reports the simplest of these analyses — a bivariate correlational analysis of the two sets of variables using the original summative scales. Clearly, the correlations are all quite strong, especially between procedural knowledge, on the one hand, and knowledge practices, on the other. But because these bivariate relationships provide only a minimal picture of the relationship between the two sets of variables, we ran a series of confirmatory factor analyses, including a hierarchical CFA with good model fit and high factor loadings (.870 for epistemic focus and .964 for knowledge practice) indicating that the two constructs can be successfully linked to a higher order latent construct. Intriguingly, the coefficient for knowledge practice was stronger than the coefficient for knowledge focus. This indicates that although the mean scores for knowledge practices are quite low (except for knowledge communication (syntax), nevertheless, the knowledge practices are slightly more central to the higher order conception of mathematical knowledge. This is an important finding, and underscores the relative importance of knowledge practice in our overall model of disciplinarity and designing instructional tasks that enhance the opportunity of students to engage in authentic knowledge work more than they currently do now.

Table 7
Correlation matrix: epistemic focus and knowledge practices, secondary 3 Mathematics, 2010

Epistemic Focus/ Knowledge Practices	Procedural Knowledge	Conceptual Knowledge	Epistemic Knowledge	Metacognitive Knowledge
Knowledge Communication: Syntax	.611**	.575**	.402**	.335**
Knowledge Validation Tasks	.588**	.477**	.322**	.310**
Knowledge Deliberation Tasks	.583**	.470**	.318**	.310**
Knowledge Representation Tasks	.675**	.569**	.375**	.364**
Knowledge Generation Tasks	.660**	.532**	.402**	.352**
Knowledge Communication Tasks: Presentation	.411**	.331**	.230**	.231**

The next step in our analysis was to run a series of structural equation models to establish the strength of the path coefficients between mathematics knowledge focus and disciplinary practices (Figure 3). The model fit statistics are exceptionally good, indicating the model explains the pattern of empirical relationships in the data very well. In addition, the structure of the pathways within the two sets of constructs remained stable internally, although the strength of the regression coefficients declined moderately for two key pathways — knowledge of communication (syntax) to knowledge representation (from .57 to .35), and from knowledge representation to knowledge generation (from .73 to .54). Critically, the pathways linking the knowledge focus scales and the knowledge practices scales are theoretically sensible and strongly consistent with our underlying model of disciplinary knowledge, particularly the pathways from procedural knowledge to knowledge communication (syntax) (.44), from conceptual knowledge to knowledge representation (.52), and from conceptual knowledge to knowledge generation (.30). The appropriateness of these pathways is reasonably self-evident: the use of syntax is dependent on procedural knowledge, and both knowledge representation and knowledge generation are forms of conceptual knowledge. The strength of these pathways underscores our contention that conceptual and procedural knowledge are the key levers that facilitate participation in knowledge practices. Two other pathways — from metacognitive knowledge to knowledge of communication (syntax) (.23), and from conceptual knowledge to knowledge deliberation (.09) were not as strong, but theoretically sensible.

An example of mathematical problem solving that exemplifies some of these relationships focusing on algebra goes as follows:

"A ball costs $5b and a pen costs $p. Susan bought 3 balls and 2 pens. How much did Susan spend altogether?"

What is expected of the student in this example is simply to multiply 5b by 3 (balls) and multiply p by 2 and add these two quantities to find the total expenditure. The answer is $(15b + 2p). Let us think of the cognitive work that needs to be done by the student in order to solve this

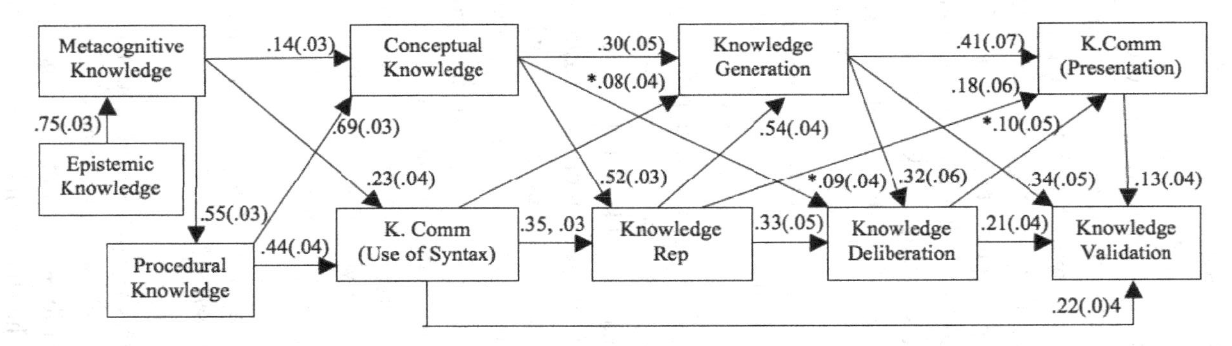

Figure 4. SEM model of knowledge focus and knowledge practices: Secondary 3 Mathematics

Note: Values on the left represent unstandardized estimates significant at $p<.01$, $*p<.05$. Values on the right are standard errors.

Goodness-of-fit statistics	
(N=1166)	
Chi-Square / d*f* / *p-value*	33.267 / 24 / .0986
CFI / TLI	.998 / .996
RMSEA (90% CI)/ SRMR	.018 (.000-.032) / .018

Note: This SEM model was constructed using composite unstandardized latent construct values rather than the original measurement values routine problem. First, the student reads and understands the question. To do so, the student engages in *knowledge communication (use of syntax). Knowledge communication* includes both receptive (in this case through reading and understanding the question) as well as expressive (in this case in giving verbal or written responses) forms of communication. Apart from an understanding of language, the correct interpretation of the word problem is afforded by a correct understanding of the syntax of mathematical language. Knowing the syntax of mathematical statements is one aspect of *procedural knowledge* (Hiebert & Lefevre, 1986). Next, the cost of three balls given the cost of one ball is calculated. This requires knowledge of the multiplication or multiple addition algorithms. Knowing algorithms or procedures in solving problems is another aspect of *procedural knowledge*.

Recall that there are two aspects of epistemic framing — *knowledge focus* and *knowledge practice. Procedural knowledge* is a part of *knowledge focus* whereas *knowledge communication (use of syntax)* is a part of *knowledge practice.* One aspect of *procedural knowledge* is knowledge of mathematical syntax, for example when we write 2 + 3 = 5, the knowledge of syntax requires us to follow the rules such as the operand "+" is placed in between the two numbers, etc. But this type of *procedural knowledge* has a *practice* counterpart in *knowledge communication (use of syntax).* This type of procedural knowledge is used when we read and understand such mathematical statements or when we write them out to communicate to other people. So in our conception of epistemic framing we make the distinction between *knowledge as something that one acquires* and *knowledge practice as something one does* in the pursuit of knowledge and in building knowledge. Inherently, our conception of *knowledge practice* has a strong bias on being a social practice but it does not preclude individuals engaged in practice. Furthermore, it allows us to see how different kind of mathematical processes are embedded in, and promoted by, different knowledge practices, thereby underscoring the role that knowledge practices play in mediating the relationship between mathematical "content" and mathematical "processes." This in turn highlights the intellectual futility of the so-called "math wars" that have so exercised generations of mathematics educators, especially in the USA.

Returning to our example, in writing $5b x 3 + $p x 2 = $15b + $2p, the student is engaged in *knowledge communication (use of correct syntax)* — correctly placing various symbols — and *representing* the quantities. In this case $15b represents the cost of 3 balls. The total cost is calculated accordingly and the student makes use of *procedural knowledge* to engage *in knowledge communication (use of syntax)* and *knowledge representation.* Now, we can see that this problem at most made use of two aspects of *knowledge practice* — *knowledge communication (use of syntax)* and *knowledge representation.* How does one transform such a task into one that requires a larger repertoire of knowledge practices?

Instead of testing whether students know how to arrive at the answer, we can ask students to *validate,* justify or substantiate different "answers". For example, $(15b + 2p); $17bp; $30bp; $25bp are some "possible" answers that students may be asked to deliberate on and explain why some answers are not possible or are wrong and to justify/validate the answer that they think is correct. The three incorrect solutions are quite common errors because beginning learners of algebra tend to be quite uncomfortable with answers that are not simplified like $(15b + 2p) and prefers a more compact solution or simple solution. At this stage other solutions may also be *generated.* $17bp, for example, can be "substantiated" by multiplying $5b by 3 to get $15b and then adding $2p to it. The next step, which is erroneous, is to add 2 to 15 making it seventeen and to collect the two algebraic symbols b and p to obtain 17bp. Indeed in generating $17bp as a solution, the answer is "substantiated" by the fact that 2 and 15 are whole numbers and separated by a + sign the "logical" step is to add these numbers to obtain 17. "b" and "p" are placed side by side for compactness thus obtaining 17bp.

The students then come to the next step of *deliberating* on how they can assess the correctness of the solution. Students need to have the *conceptual knowledge* of symbols being used here as a general value. They can then test whether, for a particular pair of values b and p, $17bp would give an answer that is correct or that is reasonable. Here, the criteria for correct or reasonable needs to be deliberated upon. Based on students' generation of different criteria set, the teacher could indicate what is meant by correct or reasonable. Yackel and Cobb (1996) discussed this issue of developing socio-mathematical norms in the classroom, i.e. what constitutes reasonable or correct in certain mathematical problems or situations. If these discussions occur within a whole class setting, or students are asked to *present* their *deliberations* and other students are allowed to ask questions or make comments, then *knowledge* is being publicly *communicated* and debated. Finally, justification or *validation* of the correct solution — $\$(15b + 2p)$ may be done using other *representations* (for example using a model) or its reasonableness indicated by numerical examples of different pairs of values of b and p resulting in the correct numerical amount. Here again, the opportunity arises for further discussion of *generation* of possible ways of *justification* through different *representations*. One interesting question can be "How many pairs of values of b and p need to be tested before one can be convinced that the "proof" is sufficient for the answer $\$(15b + 2p)$ to be correct? Or, to ask a straightforward *epistemic* question in mathematics, "what constitutes a proof?"

Let us step back and reflect on the interaction between various aspects of knowledge focus and knowledge practice. First and foremost, the goal of the task must be made clear. In the former problem, the overall goal is to acquire some *procedural knowledge*. This can be divided into sub-goals of gaining automaticity, speed and reliability in using a procedural knowledge to solve a routine problem. As noted, the work centred on *procedural knowledge, knowledge communication (use of syntax)* and *knowledge representation*. In the latter problem, the overall goal is to build knowledge. The sub-goals include acquiring *procedural knowledge*, participating in a wide range of *practices*, and developing domain specific (mathematical) or epistemic *norms* for acceptance of what can be considered knowledge.

The task examples shown above illustrate the two metaphors of learning: learning as acquisition and learning as participation (Sfard, 1998). In learning as acquisition, the focus is on acquiring some knowledge. In the problem above, this mainly is in acquiring how to calculate or represent a quantity of something in algebraic terms. In learning as participation, the focus is on the practice. Here the learning is constituted in the practice itself. While learning as acquiring knowledge traditionally is a very common goal, more recent formulations like constructivism and situated cognition are often aligned with learning as participation metaphors. But we concur with Sfard of the dangers of choosing only one metaphor. Disciplinary knowledge without the attendant processes of disciplinary practice is inert and defies the ability to apply or transfer knowledge from one context to another, whereas knowledge practice without disciplinary knowledge is epistemic nonsense. At worst, it leads to "dangerous" conclusions. Take for example, transferring the knowledge of adding two fractions as the sum of the numerators over the sum of the denominators. It is a transfer of the procedural knowledge of adding whole numbers but it leads to wrong conclusions. Why is this mistake made? Because, conceptual knowledge of what it means by a fraction is not used in applying the procedural knowledge.

Thus we advocate considering disciplinary knowledge and knowledge practice together. Our analysis has shown the intricate and dynamic relationship of disciplinary knowledge and knowledge practice in Singapore mathematics classrooms. We suggest further that this would lead to students having deep understanding as well as self-regulation of the learning processes.

5 Conclusion

We began this chapter with the general argument that the intellectual quality of knowledge work in the classroom depends substantially, although not exclusively, on the nature and quality of the tasks that teachers ask students to engage in during the course of the lesson or a unit of work. We then went on to argue that while instructional tasks have many dimensions, the epistemic framing of tasks is critical. We then specified the epistemic — as opposed to the cognitive or normative — framing of instructional tasks in terms of the epistemic (or knowledge) focus and epistemic practices associated with the knowledge work that students engage in during their mathematical lessons. We also argued that this content is not arbitrary and the practices not random, but two aspects of a historically evolving model of mathematics disciplinarity.

Finally, drawing on survey data from Secondary 3 students in a representative sample of schools across the system, we concluded that while teachers do provide students with relatively substantial opportunities to acquire procedural and metacognitive knowledge, the opportunities to acquire conceptual and epistemic knowledge are more limited. Similarly, while teachers provide relatively generous opportunities to use and understand mathematical syntax, opportunities to use and understand other forms of knowledge practice are not nearly so generous. As a consequence, teachers do not generally provide students with a rich and integrated array of opportunities to engage in knowledge practices at a level and in the kind of articulated sequence that would facilitate student understanding of the disciplinary structure of mathematics work conducted in accordance with epistemic norms sanctioned by the members of the broader mathematical community.

Beyond this, we have suggested that instructional tasks can be rendered more "authentic" — and *doing* mathematics more transparent, "visible" and accessible without being rendered more abstract or arcane — with a clearer awareness of the epistemic nature of knowledge claims and a more differentiated understanding of instructional tasks that attends to the structure and logic of knowledge practices in mathematics in ways that permit students to assume developmentally appropriate epistemic apprenticeships in the broader mathematics community of practice. Indeed, we think that there is considerable value in a stronger epistemic framing of the Singapore mathematics curriculum framework. The current curriculum formulation of concepts-skills-processes-metacognition (four sides of the pentagonal framework) in particular play into the not very sensible content-processes problematics of the kind that have characterized the "math wars" in the US (Schoenfeld, 2004). We think that an epistemic focus that encompasses content and epistemic practices with processes embedded in them make for a powerful framework that favours the dynamic interaction of knowledge and practice.

Acknowledgement

The authors want to acknowledge the contribution of the following NIE staff involved in the development of the data base used in this study: Serena Luo, Sheng Yee Zher, Khin Maung Aye, Tan Teck Kiang and Loo Siok Chen.

References

Anderson, L., & Krathwohl, D. (Eds.). (2001). *A taxonomy for learning, teaching and assessing: A revision of Bloom's taxonomy of educational objectives* (Complete ed.). New York: Longmans.

Ball, D. L. (2003). *Mathematical proficiency for all students: Towards a strategic research and development program in mathematics education.* Santa Monica: Rand.

Boaler, J. (2002a). The development of disciplinary relationships: Knowledge, practice and identity in mathematics classrooms. *Proceedings of the Annual Meeting of the International Group for the Psychology of Mathematics Education, July*, 21-26.

Boaler, J. (2002b). Exploring the nature of mathematical activity: Using theory, research and working hypotheses to broaden conceptions of mathematics knowing. *Educational Studies in Mathematics, 51*(1/2).

Brown, J. S., Collins, A., & Newman, S. E. (1989). Cognitive apprenticeship: Teaching the craft of reading, writing and matematics. In L. Resnick (Ed.), *Knowing, learning and instruction: Essays in honor of Robert Glaser* (pp.453-49). Hillsdale, NJ: Erlbaum.

Cauley, K. (1988). Construction of logical knowledge: Study of borrowing in subtraction. *Journal of Educational Psychology, 80*, 202-205.

Cobb, P., Wood, T., & Yackel, E. (1993). Discourse, mathematical thinking, and classroom practice. In E. A. Forman, N. Minick , & C. A. Stone (Eds.), *Contexts for learning: Sociocultural dynamics in children's development* (pp.91-119). New York: Oxford University Press.

Doyle, W. (1983). Academic work. *Review of educational research, 53*, 159-199.

Engle, R., & Conant, F. (2002). Guiding principles for fostering productive disciplinary engagement: Explaining an emergent argument in a community of learners classroom. *Cognition and Instruction, 20*(4), 399-483.

Engle, R., & Faux, R. (2006). Towards productive disciplinary engagement of prospective teachers in educational psychology: Comparing two methods of case based instruction. *Teaching Educational Psychology, 1*(2), 1-22.

Ford, M. J., & Forman, E. A. (2006). *Redefining disciplinary learning in classroom contexts.* American Education Research Association, Review of Educational Research. Washington: DC: In J. Green, A. Luke & G. Kelly (Eds.).

Galton, M. (2007). *Learning and teaching in the Primary classroom.* London: Sage.

Goldin, G., & Shteingold, N. (2001). Systems of representations and the development of mathematical concepts. In A. Cuoco, & F. Curcio (Eds.), *The roles of representation in school mathematics* (pp.1-23). Reston, VA: National Council of Teachers of Mathematics (NCTM).

Hattie, J. (2009). *Visible learning. A synthesis of over 800 meta-analyses relating to achievement.* London: Routledge.

Henningsen, M., & Stein, M. (1997). Mathematical tasks and student cognition: Classroom-based factors that support and inhibit high-level mathematical thinking and reasoning. *Journal of research in mathematics education, 28*(5), 524-549.

Hiebert, J., & Lefevre , P. (1986). Conceptual and procedural knowledge in mathematics: An introductory analysis. In J. Hiebert, *Conceptual and procedural knowledge: The case of mathematics* (pp.1-27.). Hillsdale, NJ: Lawrence Erlbaum Associates.

Hiebert, J., & Wearne, D. (1996). Instruction, understanding and skill in multidigit addition and instruction. *Cognition and Instruction, 14*, 251-283.

Hogan, D. (2007). Towards 'Invisible colleges': Conversation, disciplinarity and pedagogy in Singapore. *Keynote address at redesigning pedagogy conference, May 30.* Paper presented at the Centre for Pedagogy and Practice Conference, National Institute of Education: Nanyang Technological University, Singapore.

Hogan, D., Towndrow, P., Abdul Rahim, R., Chan, M., Luo, S., Sheng, Y., et al. (2011). *Interim report on pedagogical practices in Singapore in Secondary 3 mathematics and English, 2004 and 2010.* National Institute of Education, Singapore.

Kapur, M., & Bielaczyc, K. (2011). Designing for productive failure. *Journal of the Learning Sciences* (DOI:10.1080/10508406.2011.591717), 1-39.

Kaput, J. (1985). Representation and problem solving: Methodological issues related to modelling . In E. A. Silver (Ed.), *Teaching and learning problem solving: Multiple research perspectives* (pp. 381-398). Hillsdale, NJ: Erlbaum.

Kaput, J. (1987a). Representation systems and mathematics. In C. Janvier (Ed.), *Problems of representation in the teaching and learning of mathematics* (pp. 19-26). Hillsdale, NJ: Erlbaum.

Kaput, J. (1987b). Towards a theory of symbol use in mathematics. In C. Janvier (Ed.), *Problems of representation in the teaching and learning of mathematics* (pp. 159-195). Hillsdale, NJ: Erlbaum.

Lave, J. (1988). *Cognition in practice: Mind, mathematics and culture in everyday life.* Cambridge, UK: Cambridge University Press.

Lave, J., & Wenger, E. (1991). *Situated learning: Legitimate peripheral participation.* Cambridge: Cambridge University Press.

MacKinnon, D. P. (2008). *Introduction to statistical mediation analysis.* New York: Lawrence Erlbaum Associates.

McConachie, S., & Petrosky, A. (Eds.). (2010). *Content matters: A disciplinary literacy approach to improved student learning.* San Fransisco: Jossey-Bass.

NCTM. (1991). *Professional standards for teaching mathematics.* National Council of Teachers of Mathematics (NCTM), Commission on Teaching Standards for School Mathematics. Reston: Author.

NCTM. (2000). *Principles and standards for school mathematics.* National Council of Teachers of Mathematics (NCTM). Reston: Author.

Newmann, F. M., & Associates. (1996). *Authentic achievement: Restructuring schools for intellectual quality.* San Francisco, California: Jossey-Bass Publishers.

Newmann, F. M., Bryk, A., & Nagaok, J. (2001). *Authentic intellectual work and standardized tests: Conflict or coexistence.* Paper presented at the Consortium on Chicago School Research.

Putnam, R., Lampert, M., & Peterson, P. (1990). Alternative perspectives on knowing mathematics in elementary schools. *Review of research in education, 16.*

Resnick, L. B., Saljo, R., Pontecorvo, C., & Burge, B. (Eds.). (1997). *Discourse, tools and reasoning: Essays on situated cognition.* Berlin: Springer-Verlag.

Rittle-Johnson, B., & Alibali, M. W. (1999). Conceptual and procedural knowledge of mathematics: Does one lead to the other? *Journal of Educational Psychology, 91*(1), 175-189.

Sawyer, R. K. (Ed.). (2006). *Cambridge handbook of the learning sciences.* New York: Cambridge University Press.

Schoenfeld, A. H. (1992). Learning to think mathematically: Problem solving, metacognition and sense-making in mathematics. In D. Grouws (Ed.), *Handbook for research on mathematics teaching and learning* (pp. 334-370). New York: MacMillan.

Schoenfeld, A. H. (2004). The math wars. *Educational Policy, 18*(1), 253-286.

Sfard, A. (1998). On two metaphors for learning and the dangers of choosing just one. *Educational Researcher, 27*(2), 4-13.

Stein, M. K., Grover, B. W., & Henningsen, M. (1996). Building student capacity for mathematical thinking and reasoning: An analysis of mathematical tasks used in reform classrooms. *American educational research journal, 33*(2), 455-488.

Tatsuoka , K., Corter, J., & Tatsuoka, C. (2004). Patterns of diagnosed mathematical content and process skills in TIMSS-R across a sample of 20 countries. *American Educational Research Journal, 41*(4), 901-926.

Tharp, R., & Gallimore, R. (1998). *Rousing minds to life: Teaching, learning and schooling in social context.* Cambridge: Cambridge University Press.

Yackel, E., & Cobb, P. (1996). Sociomathematical norms, argumentation and autonomy in mathematics. *Journal for Research in Mathematics Education, 27*(4), 458-477.

Young, M. (2008). *Bringing knowledge back in: From social constructivism to social realism in the sociology of education.* London: Routledge.

Chapter 3

Modifying Textbook Exercises to Incorporate Reasoning and Communication into the Primary Mathematics Classroom

Denisse R. THOMPSON

Reasoning and communication are critical processes in the mathematics curriculum. As students reason and communicate about mathematics, teachers gain insight into their students' thinking that can help guide instruction. This chapter illustrates a variety of specific and general strategies that can be used to modify typical textbook exercises to incorporate reasoning and communication and make student thinking visible. Sample items and modifications are presented.

1 Introduction

Textbooks are a ubiquitous feature of classrooms around the world. As noted by Valverde, Bianchi, Wolfe, Schmidt, and Houang (2002), textbooks serve to translate educational standards from state or national policy documents into "potential pedagogical implementation strategies" (p. 167). So, it is natural to consider whether the textbooks used for instruction support the vision of mathematics curriculum of a given state or country or whether some features of textbooks are lacking, and if so, how they might be modified. Indeed, in their study of textbooks as part of the Third International Mathematics and Science Study, Valverde and his colleagues found that content changes related to curriculum reform were often integrated into the textbook, but that processes, such as problem solving or communication, were less likely to be incorporated in an effective manner.

Teachers typically use the exercises in textbooks as a source of opportunities for students to practice important content and mathematical processes (Kaur, 2010; Remillard, 1999). Such exercises are a crucial component of ongoing formative assessment, that is, assessment "designed (a) to assess where a student is in the learning process and (b) to help a teacher use students' responses to determine the instructional activities necessary to further the student's learning" (Collins, 2011, p. 9). At issue is whether textbook exercises provide sufficient opportunities for teachers to gain insights into their students' thinking in order to modify and adapt instruction to enhance student learning. This chapter focuses on ways that textbook exercises might be modified to integrate the important processes of reasoning and communication.

2 Reasoning and Communication as Essential Mathematical Processes

Students need to practice to become proficient at procedures and algorithms. But procedural fluency is just one facet of mathematical proficiency. In their review of research on mathematics learning in grades K-8, Kilpatrick, Swafford, and Findell (2001) conceptualized mathematical proficiency as consisting of five intertwined strands: conceptual understanding, procedural fluency, strategic competence (formulating and representing problems), adaptive reasoning (justification), and productive disposition. Students need many opportunities to engage with the strands other than procedural fluency if they are to become proficient with those strands as well. Indeed, the National Council of Teachers of Mathematics (NCTM, 2000) has suggested that "[r]easoning is a habit of mind, and like all habits, it must be developed through consistent use in many contexts" (p. 56).

Recommendations for curriculum reform in the United States (Common Core State Standards Initiative, 2010; NCTM, 2000) as well as in Singapore (Ministry of Education, 2006) place emphasis on important mathematical processes, including reasoning and communication. As students communicate their thinking about concepts, they not only make their learning visible to teachers but they also internalize and solidify their understanding for themselves. In addition, through students' active participation, the classroom has the potential to become more student-centered and engaging.

How often have you thought you understood a concept until you went to explain it to someone else and had difficulty articulating your thinking? The same is true for our students. By providing more opportunities for them to articulate their understanding, we help them develop robust knowledge and we gain insights that can help inform our instructional planning.

The remainder of this chapter addresses two guiding questions:

- How can we ensure that students have many opportunities to engage with reasoning and mathematical communication throughout their primary curriculum?
- How can those opportunities provide teachers with insight into students' thinking that can help teachers modify and enhance instruction?

3 Strategies for Modifying Textbook Exercises

As stated previously, the textbook is a major determinant of what happens in the classroom. As Begle (1973) noted, the textbook "is a variable that on the one hand we can manipulate and on the other hand does affect student learning" (p. 209). The aim is for teachers to look for opportunities within the textbook to engage students in reasoning and communication, and when those opportunities are not present, to modify tasks to make these processes more visible. Problems do not necessarily need to be unusual or complex in order to focus more on processes; rather, they just need to be open enough to require students to think through and justify their solutions. Throughout this section, a variety of strategies for modifying typical textbook exercises so that reasoning and communication become more prominent are illustrated and discussed.

3.1 *Reframe a basic problem by including one or more conditions*

Consider the problems in Figure 1.

Original problems: $12 + 8 =$ _____ $12 \times 8 =$ _____

Modification: Find two whole numbers whose sum is 20 and whose product is 96.

a. Explain why your numbers are correct.

b. How did you start to think about the problem?

Figure 1. Sample computational problems with modification

Primary teachers typically have students complete the two individual computations in the original problems. As neither of the original problems requires students to engage in reasoning or communication, teachers do not learn the depth of their students' knowledge or useful information that can inform future instruction.

Now consider the suggested revision. Think about how students might approach the modification and the mathematics involved. They still need to know basic computations, but they have to think about the problem in reverse. They might think about two numbers whose sum is 20. If they start with 15 and 5, they obtain a product of 75. If they try 10 and 10, they obtain a product of 100. So, they might try something in between, specifically 12 and 8. In an attempt to find a solution, students might actually complete more computations than in the original problems.

Students also have to reason about number relationships. Then they have to explain why the numbers work — focusing on showing that the numbers meet two conditions — as well as how they started the problem. Those explanations can help teachers think about ways to help students who might have initially struggled with the problem.

Note also that this modification, while still focusing on number computations, also begins to foreshadow future work with algebraic thinking. Later in their mathematical experience, students might solve the same problem using the system $x + y = 20$ and $xy = 96$.

The suggested revision is easily used throughout the primary curriculum, with increasingly larger numbers. If students have access to calculators, then the problem can be used with larger numbers for which computations would not normally be done via paper and pencil, so that the focus is increasingly on mathematical reasoning. Some students might make an initial guess using two numbers with the given sum; if the product is not the desired value, students might increase one number and decrease the other by the same amount until the given product is found. Other students might start with two numbers close to the square root of the product and then adjust as needed. Students must apply some reasoning to obtain the desired values. However, it is when they explain why their numbers work (part a) and how they thought about the problem (part b) that they have communicated their thinking and made it visible.

Figure 2 provides another possible modification to the first of the two original problems in Figure 1.

Modification: Balpreet has 20 kilograms of chocolate to package in two containers under the following conditions:

- Each container has a whole number of kilograms.
- Each container has some chocolate in it.
- No two containers have the same number of kilograms.
- All 20 kilograms of chocolate must be packaged.
  a. Find one way to distribute the chocolate.
  b. Find all possible ways to distribute the chocolate.
  c. How do you know you have found all the different ways?
  d. What happens to the answers to (a) and (b) if you change one or more of the conditions?

Figure 2. A second modification for the original problem from figure 1 (adapted from Thompson & Shultz-Ferrell, 2008)

The modification in Figure 2 can be an interesting problem for students to investigate in small groups. Real opportunities for reasoning and communication occur with parts (b) and (c). There are only 9 possibilities under the given conditions [1 and 19, 2 and 18, 3 and 17, 4 and 16, 5 and 15, 6 and 14, 7 and 13, 8 and 12, 9 and 11] before either the two containers have the same number of kilograms [10 and 10], or the values repeat.

What are some possible changes in conditions that make sense? Could the two containers have the same amount? Are there situations, such as two containers of different sizes or shapes, for which it might make sense for 3 and 18 to be a different solution than 18 and 3? Students can discuss the types of conditions they are willing to change, thereby reasoning about the use of mathematics in the real world and in the given problem. Such "what if...?" scenarios can help reinforce critical thinking as students analyze changes in the problem and in the resulting solution (Krulik & Rudnick, 1999).

As students investigate problems such as the one in Figure 2, they should be encouraged to share solutions they have found and explain why their solution is acceptable. Students should recognize that justifying solutions is a regular, on-going process in the classroom.

3.2 *Use relationships to find patterns or predict other results*

Consider the problem and related set of modifications in Figure 3. Although the original problem simply requires computation, the first modification requires students to estimate. They might think that 6.5×10 is 65, and doubling the multiplier yields an amount within the given bounds; actually any number between 17 and 21 is acceptable. Some computations, such as multiplying by 20, are easy to do mentally, if that is the goal. If students have calculators, the goal might be to find a multiplier in as few attempts as possible; students are then encouraged to think carefully about multipliers as they make multiple attempts through an iterative process.

Original. Multiply 6.5 by 5.

Modifications.

a. Multiply 6.5 by a number so the product is between 110 and 140. Record your attempts. How did you decide what numbers to try?
b. Multiply 6.5 by a number so the product is between 1100 and 1400. Explain how your work in (a) could help you do this problem in one step.
c. Multiply 0.65 by a number so the product is between 1100 and 1400. How does your work in (a) or (b) help you answer this problem?

Figure 3. Using mathematical relationships to predict results (adapted from Thompson & Shultz-Ferrell, 2008)

The real opportunities for reasoning occur as students consider parts (b) and (c), particularly if the goal is to have students find solutions by using their result in part (a) rather than start over to find a given number. In part (b), students should reason that the product has been multiplied by a factor of 10, so the multiplier from part (a) must also be increased by a factor of 10. Likewise, to relate parts (b) and (c), students might reason that the 6.5 from part (b) was decreased by a factor of 10 so that the multiplier from (b) must be increased by a factor of 10 for the product to remain unchanged.

The concept underlying the problem, multiplication of a decimal number by a whole number, has not changed in the modifications of Figure 3. However, by taking the original closed problem and modifying it slightly, students can explore and reason about important relationships among numbers (Carroll, 1999).

3.3 *Generate conjectures for students to investigate*

Teachers who have taught a particular grade for many years often see many of the same errors in their students' thinking year after year. So, typical misconceptions can be one means to form conjectures so students can reason about those concepts. Consider the original problem and modifications in Figure 4.

Original. Rewrite using the distributive property: $4(5 - x)$.

Modification 1. For all real numbers 5, x, and y, is it true that $5(x - y) = 5x - y$? Explain your reasoning.

Modification 2. For all real numbers 5, x, and y, is $5(x - y) = 5x - y$ *always* true, *sometimes* true, or *never* true? Justify your answer.

Figure 4. Restating a problem as a conjecture

Teachers know that many students fail to apply the distributive property appropriately, often writing the original problem as $20 - x$. So, rather than continue to focus on the property directly as with the original problem, the first modification gives students an opportunity to think about the problem another way. The explanation is the essential part of the question, because the explanation makes students' thinking visible.

Students should realize that one instance that does not make the sentence true disproves it, but one example that works is not a proof. Thus, through the modification teachers have an opportunity to introduce important mathematical terms such as *conjecture* and *counterexample*.

Although Modification 1 provides opportunities for students to communicate mathematically, Modification 2 may actually provide more opportunities for reasoning. If students find an instance for which the sentence is not true, they still need to reason about the problem. Is the sentence *ever* true? In this case, the sentence is true for $x = y$ or $y = 0$.

Conjectures can be written in a variety of statement forms, and do not always have to arise from student misconceptions. Consider the geometric items in Figure 5.

Original. A quadrilateral has three angles with measures 100°, 60°, and 130°. What is the measure of the 4th angle?

Modification 1. Show that a quadrilateral cannot have four obtuse angles.

Modification 2. Is it possible for a quadrilateral to have four obtuse angles? How do you know?

a. Can a triangle have more than one obtuse angle? Explain your thinking.
b. Can a quadrilateral have two obtuse angles? If so, draw a picture. If not, explain why not.
c. Can a quadrilateral have three obtuse angles? If so, draw a picture. If not, explain why not.

Figure 5. Reasoning about a geometric figure

The original problem is a closed problem; students simply add the given angle measures and subtract from 360°. Each of the two modifications provides opportunities for students to investigate the following conjecture: *A quadrilateral can have four obtuse angles.* However, the nature of the potential reasoning varies for the two modifications.

In Modification 1, students are directed on what to show. In contrast, Modification 2 is more open and students must function similarly to the way a mathematician functions, that is, test to see if the given statement might be possible and then reason accordingly. For students who need some scaffolding with such problems, the three parts to Modification 2 may generate suggestions on how to proceed.

As teachers, there are times when we want to give students explicit directions on how to approach a problem as in modification 1. But such modifications provide less opportunity for reasoning than items such as those in modification 2. Both types of reasoning are valuable and give more insight into students' thinking than the original problem. Our choice of which to use depends on our overall goal for a lesson and the purpose for the given task.

3.4 *Encourage students to solve a problem in multiple ways*

Krulik and Rudnick (1999) suggest that attempting to solve a problem using various solution strategies provides opportunities for students to engage in creative thinking. Consider the problems in Figure 6.

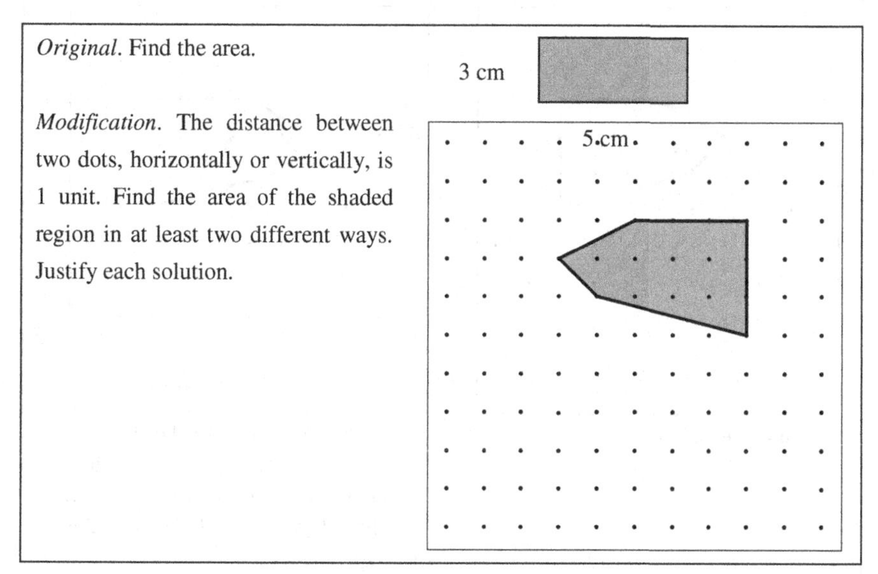

Figure 6. Modifying an area problem

The original problem suggests a basic application of the area formula for a rectangle. In contrast, the problem in the modification requires students to consider different ways that the figure might be deconstructed. Some deconstructions may be easier to evaluate than others.

Figure 7 shows two different ways that the given figure might be divided and how the area would be computed for each. The solution on the left is only shared symbolically; but the numerical sentence, together with the dashed lines in the figure, clearly indicates how the student reasoned about the problem. The student reasoned that the area of each triangular region was half the area of the related rectangle.

Likewise, the solution on the right suggests an alternative approach in which the shaded region is encased within a rectangle so that its area is found by removing the unneeded pieces from the rectangle.

Both solutions must result in the same total area. Although the focus of the modification remains on area, there are more opportunities for reasoning with the modification than with the original problem.

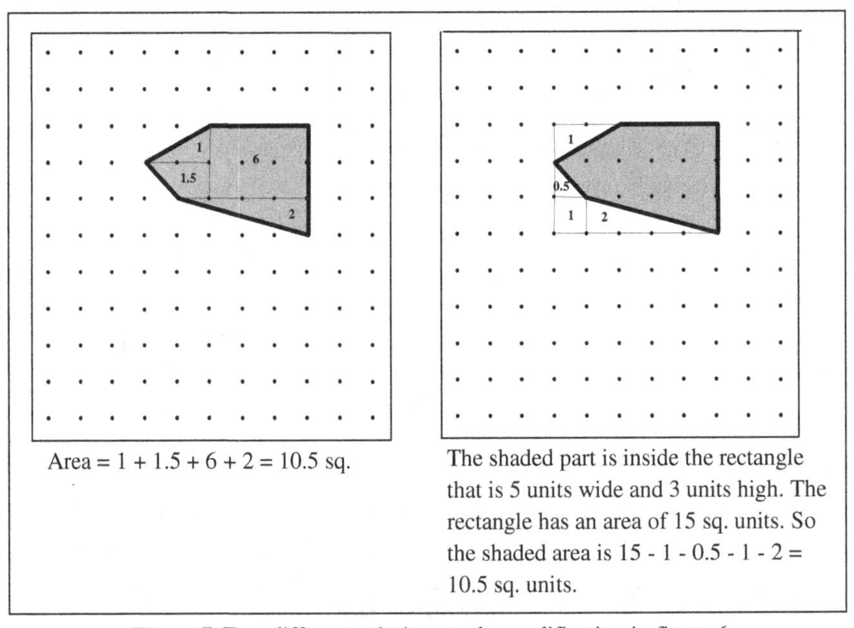

Area = 1 + 1.5 + 6 + 2 = 10.5 sq.

The shaded part is inside the rectangle that is 5 units wide and 3 units high. The rectangle has an area of 15 sq. units. So the shaded area is 15 - 1 - 0.5 - 1 - 2 = 10.5 sq. units.

Figure 7. Two different solutions to the modification in figure 6

3.5 *Evaluate student solutions*

Teachers regularly have to evaluate and validate students' solutions to problems. At times, students use an approach not anticipated by the teacher, forcing teachers to think about the problem in a different way. This activity suggests another type of modification for textbook exercises — give students sample solutions, sometimes correct and sometimes with flaws, and have them evaluate them, or have students solve a problem using another students' solution strategy.

Consider the item in Figure 8. Many students are likely to have solved the modification in Figure 6 by deconstructing the given figure into several pieces and then finding the area of each piece. Hence, the modification requires students to first carefully evaluate another students' reasoning to a problem and then to apply that reasoning to a new problem. This modification technique can work together with the technique in section 3.4 so that students develop multiple approaches to solving a problem. Teachers might also engage students in conversations about which approach is more effective or efficient in a given situation, to help students reason about the pros and cons of particular strategies.

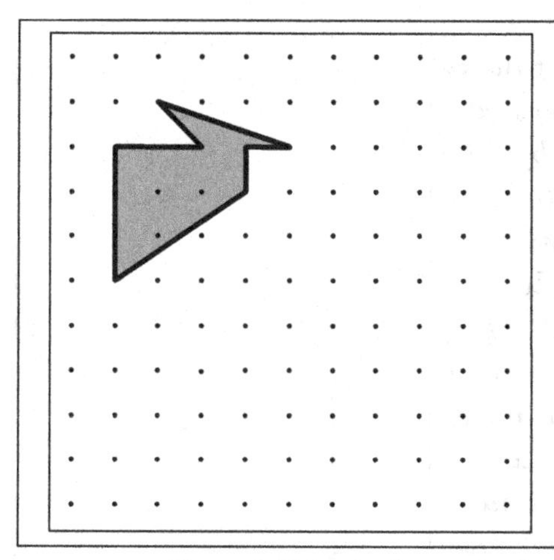

The distance between two dots, horizontally or vertically, is 1 unit. Use the solution strategy on the right in Figure 7 to find the area of the shaded region.

Figure 8. Using another student's approach to solve a problem

To address the item in Figure 8, students need to review a strategy and then apply that strategy to a new problem. In contrast, in Figure 9 students are expected to read two different student responses to a problem and evaluate them for correctness. Secondary mathematics textbooks in the United States occasionally include such problems within the exercise sets (Johnson, Thompson, & Senk, 2010), but the strategy is a powerful one for students from all levels. Students need to read and follow someone else's reasoning.

An added benefit to using such items is that students have opportunities to review and discuss what makes an acceptable solution. Problems like the one in Figure 9 are good to use for a small group activity in which students work together to critique a solution, and then share those critiques with the entire class.

Problem. Charles and David were given the following problem.

a. Give an example of a set of numbers whose mean is 100 and whose median is 150.
b. Demonstrate why your answer to part *a.* is correct.

Determine whether the responses from Charles and David are correct. Justify your answer.

David	Charles
150, 150, 25, 150, 25 25 + 25 + 150 + 150 + 150 = 500 Part a is correct because to find a mean of 100 for five numbers you would have to multiply 100 by 5 to get the missing number (500) that completes finding a mean my way. Line then up, Add, and divide. To find the median I put the 150 in the middle (median) and searched for two smaller numbers to put on the left, and two larger numbers for the right that would add up to 500 (my missing mean number).	125 + 75 + 150 + 100 + 50 = 500 ÷ 5 = 100 / mean = 100 or rather average 125, 75, (150), 100, 50 — median 125, 75, 150, 100, 50 — 1 2 (3) 4 5 — middle median 125 + 75 + 150 + 100 + 50 = 500; 5)500 = 100 = mean

Figure 9. Sample student solutions to be evaluated

3.6 *Write a question appropriate for a given answer*

Textbooks typically provide questions for which students aim to find solutions. Even with the suggested modifications that encourage reasoning and communication, students' thinking may be directed toward a particular path in order to answer the question. In contrast, having an answer and needing to generate the question requires students to think at a different level. Consider the two items in Figure 10.

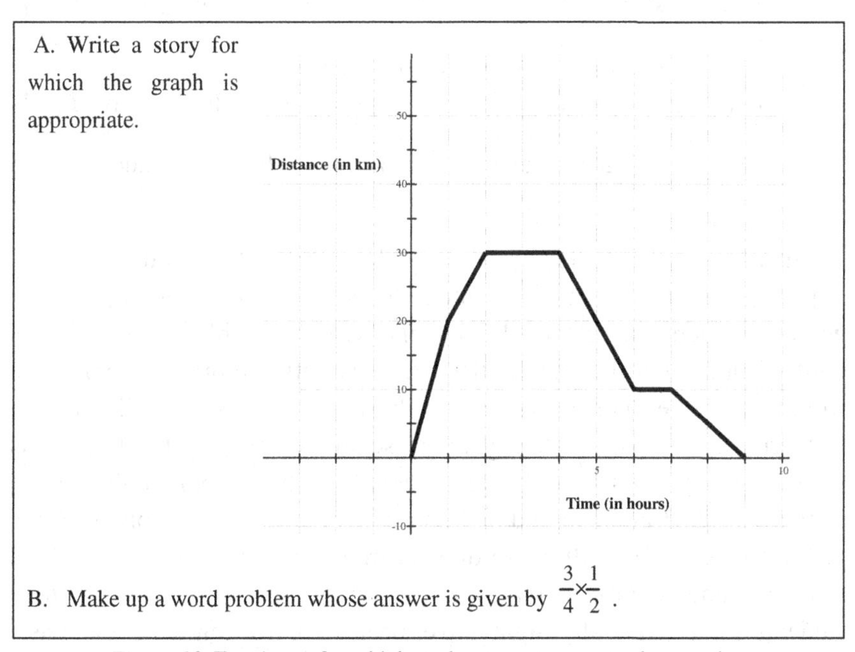

Figure 10. Two items for which students must generate the question

In item A, students must interpret the meanings of the axes and consider contexts for which the graph makes sense. Students might start out as follows: "Sally travelled from her home to the store 20 kilometers away in 1 hour. She then travelled 10 more kilometers from home in another hour." Later, when students have studied slope, they can deal with issues of speed because the slope of each segment represents distance per unit of time.

Students are accustomed to the computation involved in the second item as well as to reading a word problem and determining the needed computation. But even students who are successful solving word problems given in the textbook often struggle to write appropriate word problems on their own. Consider the following samples.

a. Mike has $\frac{3}{4}$ of a cookie and Jon has $\frac{1}{2}$ a cookie. How much do they both have together?
b. Marsha walked $\frac{3}{4}$ of a km each day. If she walked for $\frac{1}{2}$ a day every time, how many km has she walked?
c. Jane's grandmother gave her $\frac{3}{4}$ of a pie for the holiday. Jane's mother told Jane to give $\frac{1}{2}$ to her brother Joe. How much pie did Joe get?

In part a, the student wrote a problem for which addition is the required operation and used a context that would be more appropriate for whole numbers. In part b, the student wrote a problem that suggests multiplication, but there are still difficulties with the problem. Someone would not take half a day to walk three-quarters of a kilometer; furthermore, the use of "every time" seems to suggest that Marsha is walking many days, yet the given multiplication would only be the distance walked in a single day. Part c is an appropriate problem for the given computation. If all three students are in the same class, how should teachers modify instruction to address the different needs of these students? Although all might have been able to complete the basic computation or solve the textbook word problem, the inability to write one's own word problem suggests that understanding of fraction multiplication is perhaps not as robust as desired.

3.7 *Connect procedural and conceptual knowledge*

As teachers, we want students to have procedural fluency as well as conceptual understanding, two of the interconnected strands advocated by Kilpatrick, Swafford, and Findell (2001). As students connect these two strands, they have many opportunities to reason about mathematics and communicate their thinking. Consider the items in Figure 11.

In problem A, the original item simply asks students to identify a geometric figure, a task they are likely to do simply from the visual stimulus. In contrast, the modification focuses on conceptual understanding by addressing properties inherent in a rhombus.

Many students can answer the original item in B as well as generate a drawing for the given fraction. But the explanation for the picture provides important insights to teachers. Many students will draw an acceptable picture and indicate "I shaded 3 of the 8 parts." On the surface the explanation seems sufficient, but no mention is made that the pieces need to be the same size. What difficulties might the student face in the future because of this lack of precision in the meaning of a fraction?

Likewise, in problem C, students need to connect aspects of decimal addition by considering the symbolic problem from various perspectives. Some students may not be able to complete all four cells, but all students are likely to be able to engage with at least one cell. Their explanations suggest what instruction might need to be undertaken in the classroom. This task is often quite beneficial in a group setting, in which students work together to generate responses for each cell, share their work with the remainder of the class for critique, and then revise their responses as needed.

All three tasks suggest that much insight into students' thinking is possible when they connect skills with concepts. The explanations in each case once again help make student thinking visible.

A. *Original.* What shape is this?

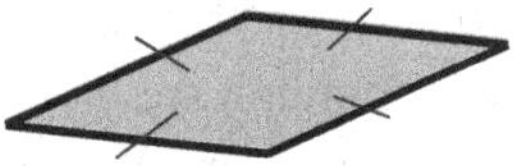

Modification. Determine the most specific figure described by the clues. Explain why you are correct. Could any clues be omitted without changing your answer? Justify.

- It has four sides.
- Opposite sides are parallel.
- All sides are congruent.
- The diagonals are perpendicular.

B. *Original.* What fraction represents the shaded part?

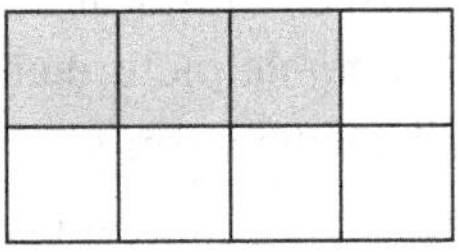

Modification. Draw a picture to show $\frac{3}{8}$. Explain how you know your picture shows the given fraction.

C. *Original.* 2.3 + 5.4 = _____

Modification. Complete each cell in the table for the topic of decimal addition (Thompson, Kersaint, Richards, Hunsader, & Rubenstein, 2008).

Mathematics Example	Real-life Example
Visual Example	Explanation in Words

Figure 11. Connecting skills with the underlying concepts

4 Conclusion

Students need many opportunities to reason and communicate about mathematics, both to enhance their own understanding of mathematics and to give teachers insight into their thinking to inform instruction. When textbook exercises do not provide sufficient opportunities with respect to these important processes, then teachers need to make adjustments in those exercises. This chapter has highlighted a number of strategies that can be used to modify typical textbook exercises.

The strategies from this chapter can be summarized and generalized as follows:

- Use vocabulary and regular questions to signal that reasoning and communication beyond a numerical answer are expected.
    - *Explain* or *Explain why.*
    - *How do you know*?
    - *Show* or *Show that.*
    - *Convince me.*
    - *Do you agree or disagree with the solution shown? Justify.*
    - *How are these problems alike or different*?
    - *How could you do this problem another way*?
- Highlight concepts that you know are potential difficulties for students.
    - Through investigating conjectures
    - Through identifying common errors
    - Through creating an argument and having students evaluate it
- Use examples of anonymous student work to generate tasks, particularly for evaluating arguments or correcting mistakes.
- Consider using language that does not give away the answer so that students must think about how to start toward a response.
    - Is it always, sometimes, or never true?
    - True or false?
    - Is the student correct? Why or why not?

Engaging students in more open tasks in which they reason or communicate does not have to be tedious, for teachers or students. Replacing one or two typical, closed tasks with more open tasks can give teachers a wealth of insight into their students' thinking. As students become convinced that such tasks are not going to disappear from the classroom and that their teachers really do value their thinking, the classroom environment can become a forum for an exchange of solution approaches and discussion. Teachers and students both win as students do more talking!

References

Begle, E. (1973). Lessons learned from SMSG. *Mathematics Teacher, 66*, 207-214.

Carroll, W. M. (1999). Using short questions to develop and assess reasoning. In L. V. Stiff & F. R. Curcio (Eds.), *Developing mathematical reasoning in grades K-12* (pp. 247-255). Reston, VA: National Council of Teachers of Mathematics.

Collins, A. M. (Ed.) (2011). *Using classroom assessment to improve student learning: Math problems aligned with NCTM and Common Core State Standards.* Reston, VA: National Council of Teachers of Mathematics.

Common Core State Standards Initiative (CCSSI). (2010). *Common Core State Standards for Mathematics.* Retrieved November 10, 2011, from http://www.corestandards.org/assets/CCSSI_Math%20Standards.pdf

Johnson, G., Thompson, D. R., & Senk, S. L. (2010). Proof-related reasoning in high school textbooks. *Mathematics Teacher, 103,* 410-417.

Kaur, B. (2010). A study of mathematical tasks from three classrooms in Singapore schools. In Y. Shimizu, B. Kaur, & D. Clarke (Eds.), *Mathematical tasks in classrooms around the world* (pp. 15-33). Rotterdam, The Netherlands: Sense Publishers.

Kilpatrick, J., Swafford, J., & Findell, B. (Eds.). (2001). *Adding it up: Helping children learn mathematics.* Washington, D.C.: National Academy Press.

Krulik, S., & Rudnick, J. A. (1999). Innovative tasks to improve critical- and creative-thinking skills. In L. V. Stiff & F. R. Curcio (Eds.), *Developing mathematical reasoning in grades K-12* (pp. 138-145). Reston, VA: National Council of Teachers of Mathematics.

Ministry of Education. (2006). *Mathematics syllabus — Primary.* Singapore: Author.

National Council of Teachers of Mathematics. (2000). *Principles and standards for school mathematics.* Reston, VA: Author.

Remillard, J. (1999). Curriculum materials in mathematics education reform: A framework for examining teachers' curriculum development. *Curriculum Inquiry, 29,* 315-342.

Thompson, D. R., Kersaint, G., Richards, J., Hunsader, P. D., & Rubenstein, R. N. (2008). *Mathematical literacy: Helping students make meaning in the middle grades.* Portsmouth, NH: Heinemann.

Thompson, D. R., & Schultz-Ferrell, K. (2008). *Introduction to reasoning and proof: Grades 6-8.* Portsmouth, NH: Heinemann.

Valverde, G. A., Bianchi, L. J., Wolfe, R. G., Schmidt, W. H., & Houang, R. T. (2002). *According to the book: Using TIMSS to investigate the translation of policy into practice through the world of textbooks.* Dordrecht, The Netherlands: Kluwer Academic Publishers.

Chapter 4

Some "What" Strategies that Advance Reasoning and Communication in Primary Mathematics Classrooms

Berinderjeet KAUR

Mathematical tasks with high cognitive demand often require students to make explicit their thinking. These tasks are necessary for the advancement of reasoning and communication in classrooms. This chapter focuses on specifically four "what" strategies that primary school teachers may use to develop such tasks to engage their students in reasoning and communication during mathematics lessons. The four "what" strategies explored in this chapter are: What number makes sense? What's wrong? What if? and What's the question if you know the answer? The chapter also features a primary one mathematics lesson during which the teacher used one of the strategies and engaged her students fruitfully in reasoning and communication.

1 Mathematical Tasks

A mathematical task is defined as a set of problems or a single complex problem that focuses students' attention on a particular mathematical idea (Stein, Grover, and Henningsen, 1996). From the TIMSS Video Study (NCES, 2003), in which Australia, Czech Republic, Hong Kong, Japan, Netherlands, Switzerland, and the United States participated, it was found that students spent over 80% of their time in mathematics class working on mathematical tasks. According to Doyle (1988), "the work students do, defined in large measure by the tasks teachers assign, determines how they think about a curricular domain and come to understand its meaning" (p. 167). Hence different kinds of tasks lead to different types of instruction, which subsequently lead to different opportunities for students' learning (Doyle, 1988). Boston and Smith (2009) report that research has consistently indicated that teachers selection of instructional tasks is largely based on lists of skills and concepts they need to cover. Textbooks are often the main source of such tasks (Doyle, 1983; Kaur, 2010).

The works of Boaler and Staples (2008), Stein and Lane (1996) and Tarr et al. (2008) have all shown that the greatest gains in student learning occur in classrooms in which mathematical tasks with high-level cognitive demand are used and the demand is consistently maintained throughout the instructional episode. Boaler and Staples (2008) in their longitudinal study comparing three high schools over a period of five years found that the highest student achievement occurred at the school in which students were supported to engage in high-level thinking and reasoning. Tarr et al. (2008) and Stein and Lane (1996) have both found that learning environments in which teachers i) encourage multiple strategies and ways of thinking; ii) support students to make conjectures and explain their reasoning, were associated with higher student performance on measures of thinking, reasoning and problem solving.

Table 1 shows a simplified version of Stein and Smith's (1998) task analysis guide that may be used to establish the cognitive demands of mathematical tasks. From the table it is apparent that tasks with high levels of cognitive demand require students to engage in explaining their thought processes.

Table 1
Levels of cognitive demand

Levels of cognitive demand	Characteristics of tasks
Level 0 — [Very Low] Memorisation tasks	- Reproduction of facts, rules, formulae - No explanations required
Level 1 — [Low] Procedural tasks without connections	- Algorithmic in nature - Focussed on producing correct answers - Typical textbook word - problems - No explanations required
Level 2— [High] Procedural tasks with connections	- Algorithmic in nature - Has a meaningful / "real-world' context - Explanations required
Level 3 — [Very High] Problem Solving / Doing Mathematics	- Non-algorithmic in nature, requires understanding of mathematical concepts and application of - Has a "real-world" context / a mathematical structure - Explanations required

Mathematics textbooks often lack tasks that are suitable for instruction to advance reasoning and communication in mathematics lessons. Therefore teachers face the challenge of crafting suitable high-level cognitive demand tasks for use in their lessons to engage their students in reasoning and communication. This challenge is not a formidable one as the works of Silver, Kilpatrick and Schlesinger (1990), Carroll (1999), Krulik and Rudnick (1999), Yeap and Kaur (1997), and Kaur and Yeap (2009a, 2009b) show that closed-ended textbook mathematical tasks can be transformed into high-level cognitive demand tasks for use in lessons to advance reasoning and communication.

Silver, Kilpatrick and Schlesinger (1990), emphasize the need to look for appropriate opportunities for thinking and communication in the material teachers are already using and are comfortable with. They suggest modifying common textbook problems to make them more open-ended as a plausible entry point for introducing speculation, group discussion and problem posing. Carroll (1999) found that one way of engaging students in the reasoning process is to have them examine and explain an error. This strategy simply involves turning closed-response questions into open-ended reasoning questions. Krulik and Rudnick (1999), in their work with mathematics teachers have also shown that standard textbook questions may be transformed into mathematical tasks capable of engaging students in critical and creative thinking, reasoning and communication (individual as well as group). They have used the following strategies: What's another way? What if? What's wrong? And What would you do? Yeap and Kaur (1997) and Kaur and Yeap (2009a, 2009b) have explored the use of problem-posing activities, drawing on typical mathematical tasks from textbooks, to promote reasoning and communication amongst students.

2 "What…" Strategies

Kaur and Yeap (2009a, 2009b) in their work with teachers in Singapore have used several "what..." strategies to engage students in reasoning and communication during mathematics lessons both in primary and secondary schools. In this chapter four of the strategies, namely: "What number makes sense?", "What's wrong?", "What if?" and "What's the question if you know the answer?" will be presented. The mathematical tasks used in these strategies are crafted from closed-ended textbook tasks.

2.1 *What number makes sense?*

In "What number makes sense?" students are presented with a mathematics version of a cloze passage that many would be familiar with in their English Language lessons. Students are presented with a problem situation which has numerical data missing. A set of numbers is provided and students determine where to place each number in the passage so that the situation makes sense. The prompts given as part of the task sheet help to focus students on the steps they need to take and also explain their thinking. The task is meant for group work. The teacher must ensure that group interaction is followed by class discussion so that students have the opportunity to explain their thinking and also learn of ways of solving problems that differ from their own. As students work through tasks of this nature, they practice computation and increase their repertoire of problem-solving skills. Reasoning skills are improved by being exposed to a variety of ways to solve a problem (Krulik and Rudnick, 2001). Such a task can be easily crafted from a typical textbook question. Figure 1 shows one such task. The mathematical task in Figure 1 was crafted from the following textbook question on the topic of numbers and number operations.

Tickets to a concert cost $15 per adult and $8 per child. Mr Wang bought tickets for 4 adults and 5 children. How much did he spend altogether?

What number makes sense?

Read the problem. Look at the numbers in the box.

Put the numbers in the blanks where you think they fit best.

Read the problem again, do the numbers make sense?

Concert tickets

Tickets to a concert cost ______ per adult and ______ per child.

Mr Wang paid ________ for _______tickets. He bought tickets

for ______ adults and _______ children.

4 5 9 $8 $15 $100

Figure 1. An example of "What number makes sense?" type of task

2.2 *"What's wrong?"*

In "What's wrong?" tasks students are presented with a problem and an erroneous solution. The error may be conceptual or computational. Students have to recognize and explain the error, why the solution was wrong and what was done to correct the error (Krulik and Rudnick, 1999). Such tasks demand higher order thinking, namely critical thinking. Students may be asked to complete the task in small groups or individually. The teacher must ensure that students are engaged in class discussion after completing the task so that they get the opportunity to see ways of solving problems that differ from their own. Furthermore, these discussions often lead to deeper mathematical understanding (Krulik and Rudnick, 2001). Teachers are in a good position to craft tasks like this as they are constantly exposed to errors students make in class and in their written assignments. Figure 2 shows one such task. The mathematical task in Figure 2 was crafted from the following textbook question on the topic of measures:

A taxi charges:

for the first 1.5 km *$2.40*

for every additional 100m *$0.10*

Mr Tan paid $12.00 for his taxi ride from work to home.

How far is Mr Tan's work place from home?

Mr Tan's trip

A taxi charges:

for the first 1.5 km $2.40

for every additional 100m $0.10

Mr Tan paid $12.00 for his taxi ride from work to home.

How far is Mr Tan's work place from home?

Raju's solution:-

$12.00 - $2.40 = $9.60

$ 9.60 ÷ $0.10 = 9.6

9.6 x 1.5 = 14.4

Mr Tan's work place is 14.4 km from his home.

There is something wrong with Raju's solution.

Show how you would solve the problem.

Explain the error in Raju's solution.

Figure 2. An example of "What's wrong" type of task

2.3 *"What if?"*

In "What if?" kind of tasks students are presented with a mathematical task following which aspects of the given information are modified, one at a time. This modification provides students with an opportunity to re-examine the task and see what effect these changes have on the solution process as well as the answer. The next part of the task requires the generation of "what if" questions by the students. This process engages students in problem posing (Brown and Walter, 1985). Such tasks are non-routine and demand higher order thinking, namely critical and creative thinking, by the students. Whole class discussion must precede individuals working on such tasks because students need to share the "what if" tasks they created with others and also make their thinking visible. Teachers are in a good position to craft tasks like this as they merely need to extend typical textbook type of questions with 'what if' conditions. Figure 3 shows one such task. The mathematical task in Figure 3 was crafted from the following textbook question on the topic fractions.

Mrs Chen bought 2 kg of flour. She used 750 g for baking cakes.
i) What fraction of the flour did she use?
ii) What fraction of the flour was left?

Baking cakes

Mrs Chen bought 2 kg of flour. She used 750 g for baking cakes.

i) What fraction of the flour did she use?

ii) What fraction of the flour was left?

What if Mrs Chen bought 3 kg of flour?
What if Mrs Chen used 900 g for baking cakes?

Generate another 3 "What if" tasks and answer them.
Look out for any interesting observations / patterns.

Figure 3. An example of a "What if?" type of task

2.4 *"What's the question if you know the answer?"*

In "What's the question if you know the answer?" kind of tasks students are presented with the context and data but with the question/s missing. Students are asked to write a question that matches a given answer. Such tasks engage students in critical thinking. Whole class discussion must precede individuals working on such tasks as it is important for students to learn that several questions may have the same answer, but certainly different solutions. Figure 4 shows an example of one such task. The mathematical task in Figure 4 was crafted from the following textbook question on the topic money and measures.

> *A rectangular piece of carpet is placed on the floor of a room leaving a margin 1 metre around it. The room measures 7 m by 6 m. Find the cost of the carpet if 1 m^2 of it costs \$75.*

The carpet

A rectangular piece of carpet is placed on the floor of a room leaving a margin 1 metre around it. The room measures 7 m by 6m. The carpet costs \$75 per m^2.

1. What's the question if the answer is 42 m^2?
2. What's the question if the answer is 20 m^2?
3. What's the question if the answer is 22 m^2?
4. What's the question if the answer is \$1500?

Figure 4. "What's the question if you know the answer" type of task

3 A Primary One Mathematics Lesson

In this section, we feature a primary one mathematics lesson during which the teacher used the strategy "What's the question if you know the answer" and infused reasoning and communication in her lesson. The teacher acquired the strategy during her participation in a project, the EPMT (Enhancing the Pedagogy of Mathematics Teachers) (Kaur, 2009; 2011), conducted by the author and her colleague.

3.1 *Objectives of the tasks*

Figures 5 and 6 show the tasks that the teacher designed and enacted in her lesson for primary 1 students. The objectives provided by the teacher for the use of tasks X1 and X2 in her lesson were
i) to provide students with an opportunity to reason and communicate,
ii) to review numbers less than 40 and the four operations, +, −, × and ÷.

Eggs-cellent?

Mrs Lee has a farm. She collects eggs from her farm every day.

Day	Number of eggs collected
Monday	23
Tuesday	17
Wednesday	8
Thursday	5
Friday	10

What's the question if the answer is 5?
What's the question if the answer is 25?
What's the question if the answer is 6?
What's the question if the answer is 30?

Figure 5. Task X1

Old MacDonald had a Farm
There are 5 cows and 12 ducks at a farm.
Each cow has 4 legs and each duck has 2 legs.

What is the question if the answer is 17?
What is the question if the answer is 7?
What is the question if the answer is 20?
What is the question if the answer is 44?

Figure 6. Task X2

3.2 *How the tasks were enacted*

This 60 minutes lesson was a review of numbers less than 40 and the four operations (addition, subtraction, multiplication and division). The lesson began with the teacher reviewing number sentences using numbers less than 40 and the four operations. Next the teacher engaged the whole class step by step in doing task X1. The students sat on the floor at the front of the whiteboard in the class which displayed task X1. The teacher explained to the students what the task was all about and very carefully led them to craft questions that resulted in the answers given. This was followed by seatwork that adopted the Think-Pair-Share strategy. Students worked in pairs on task X2, while the teacher walked around the class providing guidance. Teacher also gathered common difficulties the students faced and drew the attention of the whole class to them and scaffold their thinking. Teacher also monitored the progress of students' work and facilitated the sharing amongst the students in groups. As there was insufficient time for the students to work through all the items in task X2, the teacher limited classwork to only the first two. 50 minutes into the lesson, the teacher drew the attention of the whole class to the diversity of questions the students had crafted and also some very creative ones. The lesson ended with the teacher assigning the students to do the last two items of task X2 as homework.

3.3 *Teacher's self-evaluation of her lesson*

The teacher was guided in the self-evaluation of her lesson by the following prompts:

i) Was this lesson different from lessons in the past that you conducted? If so, what enabled you to make it different?
Teacher's response: In the past, I reviewed numbers less than 40 and the four operations by doing whole class oral work and individual written work. My students did exercises like: $7 + 9 = ?$, $12 - 5 = ?$, $5 \times 4 = ?$, $10 \div 2 = ?$ They also made number sentences with three numbers, like 7, 8, 15. This lesson is different. I used the strategy "what is the question if the answer is?" that I learned in the project. I want my students to make number sentences, but only one number is given, the answer. They must reason and pick the other two numbers from the data provided. I want them to discuss with their classmates. I also want them to see many possibilities their friends may have different number sentences. I want them to articulate their questions. They always read questions in their textbooks. This is their chance to put the questions together. The project has introduced me to many good strategies and I like to use as many of them as possible in my lessons.

ii) What were the challenges you faced in executing the lesson plan?
Teacher's response: The biggest challenge was to maintain the cognitive demand of the task, i.e. not to take over the thinking that I wanted my students to do. So, for the whole lesson, I answered all the queries of my students with appropriate counter questions, such as "what do you think?", "how would you check", etc. I also over planned my lesson. There was not enough time towards the end of the lesson for presenting all the diverse and creative questions the students had formed.

iii) Would you use the strategy again in future lessons?
Teacher's response: My students enjoyed the lesson very much. They got lots of opportunity to talk. The diverse responses of their classmates introduced them to look for more than one way. I will use this strategy again in my lessons but the purpose of my lesson will decide when and for what topic I may use it. I personally find this strategy useful for review lessons.

4 Conclusion

The four "What" strategies described in this chapter provide teachers with ways to advance reasoning and communication in their lessons whilst doing mathematics which could have otherwise been done through expository methods, such as telling and showing followed by practice. The lesson featured in this chapter is deliberate to show that students are never too young or the mathematics content knowledge too basic for students to engage in meaningful reasoning and communication during mathematics lessons.

Acknowledgement

The author is grateful to the teacher and her students whose lesson is featured in this chapter for their contribution.

References

Boaler, J., & Staples, M. (2008). Creating mathematical futures through an equitable teaching approach: The case of Railside School. *Teachers College Record, 110*, 608-645.

Boston, M.D. & Smith, M.S. (2009). Transforming secondary mathematics teaching: Increasing the cognitive demands of instructional tasks used in teachers' classrooms. *Journal for Research in Mathematics Education, 40*(2), 119-156.

Brown, S. I., & Walter, M. I. (1985). *The art of problem posing*. Philadelphia, PA: Franklin Institute Press.

Carroll, W. M. (1999). Using short questions to develop and assess reasoning. In L. Stiff (Ed.), *Developing mathematical reasoning in grades K-12* (pp. 247-255). VA, Reston: National Council of Teachers of Mathematics.

Doyle, W. (1983). Academic work. *Review of Educational Research, 53*, 159-199.

Doyle, W. (1988). Work in mathematics classes: The context of students' thinking during instruction. *Educational Psychologist, 23*, 167- 180.

Kaur, B. (2009). Enhancing the pedagogy of mathematics teachers (EPMT): An innovative professional development project for engaged learning. *The Mathematics Educator, 12*(1), 33-48.

Kaur, B. (2010). A study of mathematical tasks from three classrooms in Singapore schools. In Y. Shimizu, B. Kaur & D. Clarke (Eds.), *Mathematical tasks in classrooms around the world* (pp. 15-33). Rotterdam: Sense Publishers.

Kaur, B. (2011). Enhancing the pedagogy of mathematics teachers (EPMT) project: A hybrid model of professional development. *ZDM — The International Journal on Mathematics Education, 43*(6/7), 00-00.

Kaur, B. & Yeap, B.H. (2009a). *Pathways to reasoning and communication in the primary school mathematics classroom*. Singapore: National Institute of Education.

Kaur, B. & Yeap, B.H. (2009b). *Pathways to reasoning and communication in the secondary school mathematics classroom*. Singapore: National Institute of Education.

Krulik, S. & Rudnick, J.A. (1999). Innovative tasks to improve critical and creative thinking skills. In L. Stiff (Ed.), *Developing mathematical reasoning in grades K-12* (pp. 138- 145). Reston, VA: National Council of Teachers of Mathematics.

Krulik, S. & Rudnick, J.A. (2001). *Roads to reasoning — Developing thinking skills through problem solving* [Grades 1-8]. Chicago, IL: Wright Group McGraw-Hill.

NCES (National Center for Educational Statistics). (2003). *Teaching mathematics in seven countries: Results from the TIMSS video study*. Washington, DC: U.S. Department of Education.

Silver, E.A., Kilpatrick, J. & Schlesinger, B. (1990). *Thinking through mathematics — Fostering inquiry and communication in mathematics classrooms*. College Entrance Examination Board.

Stein, M. K., & Smith, M.S. (1998). Mathematical tasks as a framework for reflection: From research to practice. *Mathematics Teaching in the Middle School, 3*(4), 268-275.

Stein, M. K., Grover, B., & Henningsen, M. (1996). Building student capacity for mathematical thinking and reasoning: An analysis of mathematical tasks used in reform classrooms. *American Educational Research Journal, 33*, 455-488.

Stein, M.K., & Lane, S. (1996). Instructional tasks and the development of student capacity to think and reason: An analysis of the relationship between teaching and learning in reform mathematics project. *Educational Research and Evaluation, 2*, 50-80.

Tarr, J.E., Reys, R.E., Reys, B.J., Chavez, O., Shih, J., & Osterlind, S. (2008). The impact of middles-grades mathematics curricula and the classroom learning environment on student achievement. *Journal for Research in Mathematics Education, 39*, 247-280.

Yeap, B. H. & Kaur, B. (1997). Problem posing to promote mathematical thinking. *Teaching and Learning, 18*(1), 64-72. Singapore: National Institute of Education.

Chapter 5

Reasoning and Justification in the Secondary Mathematics Classroom

Denisse R. THOMPSON

Reasoning and justification are critical processes in the mathematics curriculum, yet students often struggle with these processes. This chapter uses a framework from textbook research to illustrate types of exercises that can be used to enhance the extent to which reasoning and justification are evident in the curriculum. Typical items from the algebra and statistics portion of the secondary curriculum are shared, with suggestions for modifications that highlight reasoning and justification.

1 Introduction

Reasoning, and the communication about mathematical understanding that reasoning entails, should be a critical part of the secondary mathematics curriculum. In the *Principles and Standards for School Mathematics* (National Council of Teachers of Mathematics [NCTM], 2000), reasoning is identified for the U.S. curriculum as a critical "habit of mind, and like all habits, it must be developed through consistent use in many contexts" (p. 56). Curriculum documents in other countries, such as Singapore, have also identified reasoning as a critical mathematical process (Ministry of Education, 2006) with which students need to engage.

At the secondary level, reasoning has long been an explicit part of the geometry curriculum (Herbst, 2002; Sinclair, 2008), but is sometimes less explicit in other parts of the curriculum. However, given the fundamental importance of reasoning and proof to students' mathematical understanding, it is essential that these processes be integrated throughout the secondary mathematics curriculum. Indeed, through proof, students can

- "explain why a particular mathematical result must be true,
- ... [become] autonomous learners ... [with the skills] to evaluate the validity of their own reasoning and that of others, ...
- [make] connections and provide insight into the underlying structure of mathematics." (Knuth, 2000 as cited in NCTM, 2009, p. 5)

Despite the fact that numerous educators view reasoning as critical to both *doing* and *learning* mathematics (e.g., Epp, 1998; Harel & Sowder, 2007; Herbst & Brach, 2006), students face numerous difficulties with proof. Some of the difficulties identified by researchers across many studies include the following:

- Recognizing the purpose of a proof;
- Accepting empirical examples as sufficient for a proof;
- Lacking knowledge of definitions, notation, or essential concepts needed in a proof;
- Unfamiliarity with different proof strategies or with how to start a proof;
- Being able to monitor one's own progress while working toward a proof.

For further summaries of proof-related research, readers should refer to the reviews by Harel and Sowder (2007), Yackel and Hanna (2003) and the research reported in Stylianou, Blanton, and Knuth (2009).

2 Importance of the Textbook in Providing Opportunities for Reasoning

Textbooks provide important guidance to teachers about the educational expectations for their students, both in terms of curriculum content and instructional strategies (Valverde, Bianchi, Wolfe, Schmidt, & Houang, 2002). That is, textbooks highlight *potential* opportunities students have for learning. As Begle (1973) has noted, the textbook "is a variable that on the one hand we can manipulate and on the other hand does affect student learning" (p. 209). At issue is whether secondary mathematics textbooks provide sufficient opportunities for students to engage with reasoning.

Thompson, Senk, and Johnson (in press) analyzed opportunities for proof-related reasoning in the areas of exponents, logarithms, and polynomials across 20 U.S. textbooks comprising six textbook series. After analyzing almost 10,000 exercises, they found that 5-6% of the exercises explicitly gave students opportunities to engage with reasoning in these three content areas, and it is highly unlikely that teachers would have assigned all of those exercises for students to complete. Although developing an argument and investigating conjectures were the most common types of reasoning exercises, the relatively small percentage of exercises highlighting reasoning suggests that students had somewhat limited opportunities to focus on this important mathematical process.

As students engage with reasoning and communicate their understanding, they make their learning visible both for themselves and their teachers. Teachers are then able to modify and adapt instruction to enhance student learning. One issue is the extent to which the textbook supports teachers in making reasoning an integral part of the curriculum. As they use their textbooks to guide instruction, teachers need to recognize potential opportunities for reasoning within the curriculum; when those opportunities are lacking, they need to modify questions so that reasoning and thinking are explicitly encouraged. Thus, the remainder of this chapter focuses on the following question:

- How can we modify textbook questions to ensure that students have many opportunities to engage with reasoning, proof, and justification throughout their secondary curriculum?

3 Aspects of Reasoning to Incorporate into the Curriculum

How likely is it that you would find in mathematics textbooks proof-related reasoning as an explicit focus in more than 6% of its exercises? What aspects of reasoning are commonly present? What aspects of reasoning seem to be missing?

As part of their textbook analysis, Thompson, Senk, and Johnson (in press) developed a framework with six components of proof-related reasoning. Their framework included not only developing mathematical arguments, but components such as finding counterexamples or investigating conjectures that may serve as initial steps in the proof process. Because the textbook often serves as the major source of instruction for the mathematics classroom, it seems natural to start with the resources within the textbook and modify them as needed to integrate more reasoning. The six components of the framework developed by Thompson, Senk, and Johnson (in press) serve as the basis for easily adapting textbook exercises to enhance *potential* opportunities for reasoning. (See also Johnson, Thompson, & Senk (2010) for additional examples of each component.)

3.1 *Finding counterexamples*

As Epp (1998) has argued, finding counterexamples can be a good first step in the reasoning process because finding a counterexample is typically easier for students than writing an actual proof. Furthermore, being able to use examples and non-examples appropriately is an essential prerequisite for students to progress in their reasoning ability as they make or investigate conjectures.

As students reason about counterexamples, they need to make at least one important distinction about empirical examples within the reasoning process. That is, they need to recognize that one counterexample is sufficient to *disprove* a statement, but many examples cannot *prove* a generalization is true. For some students, a counterexample is simply an exception to the rule, particularly if there are many other examples for which the statement is true.

Consider the items in Figure 1. The original item focuses solely on algebraic manipulation. Many students are likely to expand correctly; others are likely to omit the middle term in the expansion. Although there is a need for such manipulations in the curriculum, at times such items might be modified to engage students with reasoning. Notice that the modification focuses on a typical error that students make. Think about how students might proceed to answer the modification.

Original:	Expand $(x + 4)^2$.
Modification:	Find a counterexample to show that $(x + 4)^2 \neq x^2 + 16$.

Figure 1. Modifying an algebraic manipulation to find counterexamples

For students who know the meaning of the term *counterexample*, the directions provide a clue about how to proceed, namely find an example for which $(x + 4)^2$ and $x^2 + 16$ yield different values. If students first try $x = 0$, they find that the two expressions are equivalent, so additional examples need to be tried. Indeed, any $x \neq 0$ can serve as a counterexample.

The modification provides an opportunity for an important discussion in the classroom. Certain numbers, such as $x = 0, \pm 1$, or ± 2, are not necessarily the best values to use when evaluating statements because they are too special. Students need to realize that if such values make a statement true, other values also need to be tried. If students have access to graphing calculators, graphing $y = (x + 4)^2$ and $y = x^2 + 16$ on the same coordinate axes highlights that the two expressions are equivalent only for $x = 0$.

3.2 *Investigating conjectures*

When finding a counterexample, the student knows the final result that is needed. In contrast, when investigating conjectures, students do not know if the conjecture is true or false. Thus, they must function similar to the way a mathematician functions. It is natural to try several examples. If the student finds an example for which the statement is not true, then the student has disproved the conjecture. However, just because the student finds several examples for which the statement is true does not mean the conjecture is *always* true; a more formal proof is still needed.

The item in Figure 1 could have been modified to focus on investigating conjectures. Consider the two modifications in Figure 2. For Modification 1, reasoning comes into play when trying to convince someone about the answer; students might find an example that does not make the two sides of the equation equivalent or show the two graphs.

Modification 2 may actually provide more opportunities for reasoning than Modification 1 because students must consider all three possibilities. If students try a number, such as $x = 3$, they find $7^2 \neq 25$; thus, they only know that the statement is <u>not</u> always true. They still need to determine if it is *sometimes* true.

Modification 1:	Is it true that $(x + 4)^2 = x^2 + 16$? How would you convince someone about your answer?
Modification 2:	Is $(x + 4)^2 = x^2 + 16$ *always* true, *sometimes* true, or *never* true? Explain.

Figure 2. Modifications to the Figure 1 item to investigate conjectures

It is through explanations that students make their thinking visible to teachers so that instruction can be modified as needed. Consider Figure 3 with two student responses to an item similar to the one in Figure 2. Even though both students recognize that the conjecture is false, only the first student response is an acceptable justification. In the second response, the student makes computational errors with both attempts at using specific numbers to evaluate the statement.

The point here is that the students' explanations make their thinking visible. The second student knows how to reason about conjectures but has major difficulties with algebraic computation. By reflecting on this student's work, the teacher has insights into what instruction might be needed to move the student's achievement forward.

Original:	For all numbers x and y, is it true that $x^2 + y^2 = (x + y)^2$? Explain.
Response 1:	No. Just take, for example $x = 8$ and $y = 6$. So $8^2 + 6^2 = 100$ and $(8 + 6)^2 = 196$. So, it's wrong to say "all numbers."
Response 2:	No. Show any two in here. $5^2 + 6^2 = (5 + 6)^2$ $25 + 36 = (25 + 36)^2$ $61 \neq 61^2$ $4^2 + 8^2 = (4 + 8)^2$ $16 + 64 \neq 12^2 = 24$

Figure 3. Sample responses to investigating a conjecture

3.3 *Making conjectures*

As today's classrooms become more student-centered, teachers often design lessons in a guided-discovery format so that students have opportunities to explore concepts and look for patterns. Students need to be encouraged to generalize those patterns by making conjectures whenever appropriate. Although making conjectures from numerous examples involves *inductive reasoning* and teachers do not want students to consider the use of such examples as a proof, inductive reasoning that leads to conjectures is a critical component of mathematics. In fact, the structure of the patterns observed in order to make a conjecture may suggest an approach that can later be used to develop a general mathematical argument, that is, a proof.

Consider the items in Figure 4. As students engage in finding the mean and standard deviation of data, they can also engage in reasoning about how changes to data influence the mean or the standard deviation. Often, the following theorem or generalization is simply stated in a textbook or during instruction:

- When a number k is added to each value in a data set, the mean increases by k while the standard deviation remains unchanged.

Examples may follow the statement to illustrate its meaning. Nevertheless, by starting with the statement rather than having students explore the concept as in the modification, teachers lose a valuable opportunity to engage their students in reasoning.

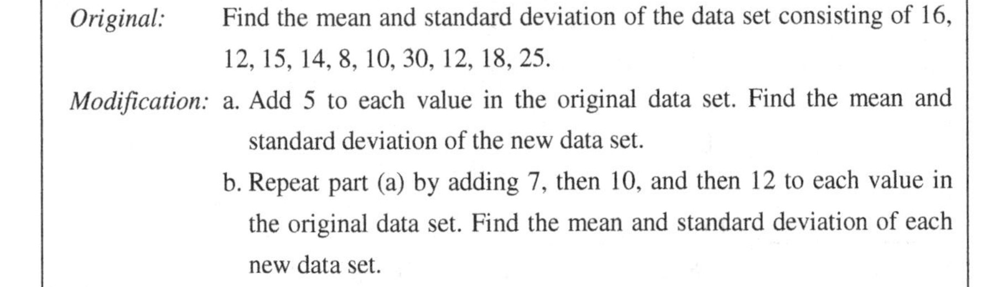

Original: Find the mean and standard deviation of the data set consisting of 16, 12, 15, 14, 8, 10, 30, 12, 18, 25.

Modification:
a. Add 5 to each value in the original data set. Find the mean and standard deviation of the new data set.
b. Repeat part (a) by adding 7, then 10, and then 12 to each value in the original data set. Find the mean and standard deviation of each new data set.
c. Make a conjecture about what happens to the mean and standard deviation when k is added to each value in a data set.

Figure 4. Modifying an item to require students to make conjectures

Students might naturally wonder whether changes to a data set in a consistent manner, either adding or multiplying by a constant, always lead to the same change in the mean but no change in the standard deviation. Given that students have already explored adding k to each value, teachers could use students' curiosity to have them explore and develop a conjecture about the effect of multiplying each value in the data set by k. Students should find that the mean and the standard deviation in this case are both multiplied by k. If students have studied transformations, they can make connections between statistics and transformations of functions, noting that adding k is similar to a translation of the data but multiplying by k is similar to a dilation of the data.

When students explore and make their own conjectures about important concepts, they are more likely to make connections and internalize the statements than when they have simply been given to them. To illustrate the power of having students develop their own generalizations, I share an experience from teaching a geometry course to prospective secondary mathematics teachers. In the course, we had previously discussed and developed a proof of the Pythagorean theorem.

When discussing the development of the distance formula in the coordinate plane, students were led through an activity to determine the distance along a horizontal segment, along a vertical segment, and then to use those results to conjecture a formula for the distance along an oblique segment by using the Pythagorean theorem. At the end of the activity, one particularly good student was quite excited. Although she had previously studied the distance formula, she had never made the connection to the Pythagorean theorem. The formula had simply been given, and she always struggled to memorize it correctly. By making her own generalization, she made an important connection that she is unlikely to forget; if she does forget the distance formula, she now has the tools to reconstruct that formula quickly. Such is the potential power of engaging students in making and developing their own conjectures for important concepts whenever possible.

3.4 *Developing arguments*

As students become accustomed to making and investigating conjectures, it becomes natural to extend their experience to developing arguments, that is, proofs. Afterall, proof is the essence of mathematics. "The possibility of proof is what makes mathematics what it is, what distinguishes it from other varieties of human thought" (Hersh, 2009, p. 17). Thus, it seems essential to engage students in this process. Indeed, "Students cannot be said to have learned mathematics, or even *about* mathematics, unless they have learned what a proof is" (Hanna, 2000, p. 24).

Conjectures that students have either investigated or made on their own can be a good source of items for which students are expected to develop an argument. For instance, after developing a conjecture for part (c) of the modification in Figure 4, students could be asked to write a general argument. Having worked through the specific cases in parts (a) and (b), students should be able to use the same techniques to write a general proof as shown in Figure 5. Writing a general argument, however, is important to ensure that students do not view many examples as sufficient for a proof.

Original: A data set has n values. If k is added to each value, show that the mean of the new data set is k more than the mean of the original data set.

Proof: Let the data values be $x_1, x_2, x_3, \ldots, x_n$. When k is added to each data value, the mean is given by

$$\bar{x}_{new} = \frac{(x_1+k)+(x_2+k)+(x_3+k)+\ldots+(x_n+k)}{n}$$

$$= \frac{(x_1+x_2+x_3+\ldots+x_n)+n \bullet k}{n}$$

$$= \bar{x}_{original} + k$$

Figure 5. Proof of the conjecture from part (c) of Figure 4

Although the ultimate goal is to have students develop general arguments, many students struggle with knowing how to begin a proof. So, teachers need to consider what strategies might be appropriate to help students overcome their initial struggles. In their analysis of secondary textbooks, Thompson, Senk, and Johnson (in press) identified numerous instances of what they called *specific arguments* and what other researchers (e.g., Movshovitz-Hadar, 1998) have called *transparent pseudo-proofs*. Consider the items in Figure 6. How do the items build on each other so that students can easily write a general proof for Item 3?

Item 1: Write an argument to show that $3^4 \bullet 3^5 = 3^9$.

Item 2: For any $x \neq 0$, show that $x^4 \bullet x^5 = x^9$.

Item 3: For all $x \neq 0$ and all positive integers a and b, show that $x^a \bullet x^b = x^{a+b}$.

Figure 6. Using specific proofs to scaffold student thinking for a general proof

Although writing an argument for properties of exponents is not necessarily difficult for students, the principle highlighted in Figure 6 is one that can be used with more difficult concepts or with students who are struggling. Clearly, a *proof* for Item 1 is a proof of a very specific case, with a specific base and specific exponents. Although the proof for Item 2 is also about a specific case, it begins to move toward generality because the base can be any non-zero number. Finally, the proof for Item 3 is a general proof of the Product of Powers Property. The arguments for Items 1 and 2 are not proofs in the typical sense, but as Movshovitz-

Hadar (1998) notes, "one can see the general proof... because nothing specific to the case enters the proof" (p. 19). Hence, these arguments about specific cases are often more concrete for students and can be used as precursors to model the structure of a general argument.

There are many opportunities within the secondary mathematics curriculum for students to develop arguments about specific cases. Students' responses make their thinking visible so that teachers can modify and adapt instruction as needed. Consider the statistics item and sample responses in Figure 7.

Item:	On one chemistry test, Amelia scored 97 when the class mean was 85 with a 4.8 standard deviation. On a second chemistry test, Amelia scored 82 when the class mean was 75 with a 2.7 standard deviation. On which test did Amelia score better in relation to the rest of the class? Write an argument to explain your thinking.
Response 1:	She scored better on the second test in relation to the class because she was about 2.6 standard deviations above average where on the first she was only 2.5. 1st test: 97 – 85 = 12, 12 ÷ 4.8 = 2.5 standard deviations above average 2nd test: 82 – 75 = 7, 7 ÷ 2.7 = 2.59259 standard deviations above average
Response 2:	On the first test it looks like Amelia did better but since the s.d. was large there were a more variety of scores. On the second test she did significantly better than her classmates and the s.d. was small meaning the majority of people were around the mean. So comparing both tests she did better on the second test compared to the class because of the smaller s.d.
Response 3:	The first test she scored better because her standard deviation was higher than on her second. Also, the first test she scored 12 more points than the rest of her class as to her 2nd test she only scored 7 more.

Figure 7. Sample item with three student arguments

Response 1 indicates a solid understanding of the strategies needed to determine the relative placement of two scores. Response 2 reflects appropriate understanding of the meaning of standard deviation; further, the student makes comparisons of a score to the mean in conjunction with the meaning of standard deviation as the variability within the score distribution. However, the student appears to make these comparisons in a global fashion from a "sense" of the numbers involved rather than providing specific documentation that the score on the second test is farther from the majority of scores in terms of standard deviation. In contrast, Response 3 suggests a misunderstanding of the meaning of standard deviation. These student responses highlight understandings and misconceptions about data in ways that simply finding the number of standard deviations from the mean would not. Teachers can then use that information to modify instruction to enhance student learning.

3.5 *Evaluating arguments*

One method that can be used to help students improve their ability to write complete, coherent arguments is to have them evaluate sample arguments from other students. Students often have their own approach to a problem. "However, evaluating the validity of someone else's proof often requires assessing the appropriateness of an approach not previously considered. Requiring students to explain how or why they determine that a proof is valid or not provides different insights into their thought processes" (Thompson & Senk, 1993, p. 170).

Reading and evaluating arguments written by others is often a good task for students to explore in small groups. As students discuss an argument, they are typically able to identify weaknesses in an argument or identify where more clarity is needed. When using this strategy with students, they are generally critical in a constructive manner, and teachers simply need to serve as facilitators of the discussion (Thompson, Beneteau, Kersaint, & Bleiler, in press). The insights students gain through such discussions about what makes a good argument are then applied to future arguments they make on their own.

Consider the item in Figure 8.

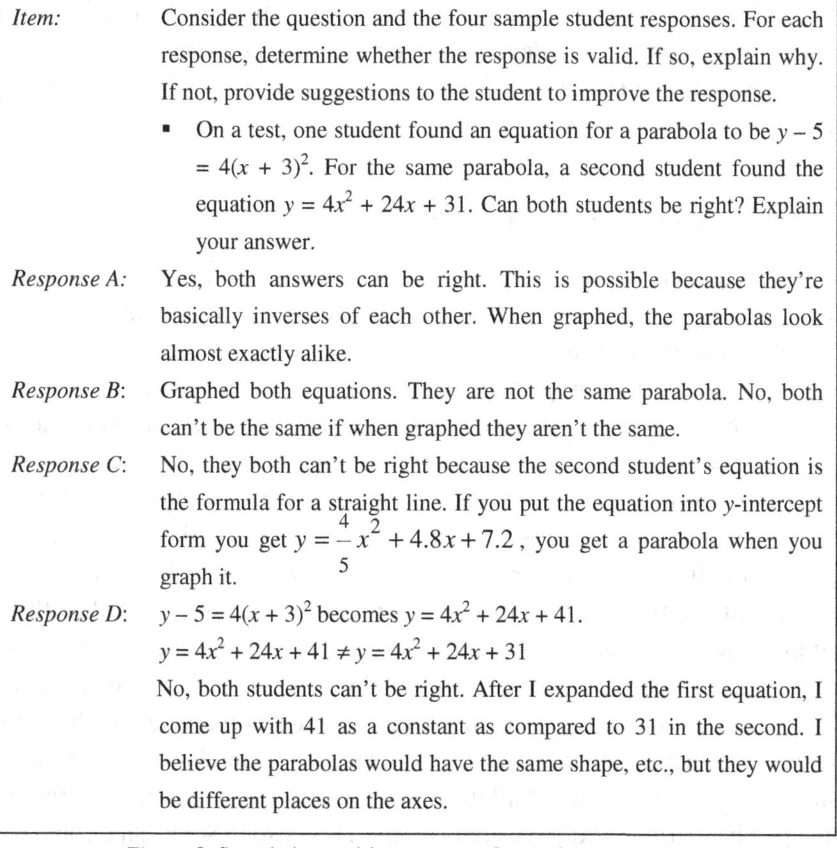

Item:	Consider the question and the four sample student responses. For each response, determine whether the response is valid. If so, explain why. If not, provide suggestions to the student to improve the response. ▪ On a test, one student found an equation for a parabola to be $y - 5 = 4(x + 3)^2$. For the same parabola, a second student found the equation $y = 4x^2 + 24x + 31$. Can both students be right? Explain your answer.
Response A:	Yes, both answers can be right. This is possible because they're basically inverses of each other. When graphed, the parabolas look almost exactly alike.
Response B:	Graphed both equations. They are not the same parabola. No, both can't be the same if when graphed they aren't the same.
Response C:	No, they both can't be right because the second student's equation is the formula for a straight line. If you put the equation into *y*-intercept form you get $y = \frac{4}{5}x^2 + 4.8x + 7.2$, you get a parabola when you graph it.
Response D:	$y - 5 = 4(x + 3)^2$ becomes $y = 4x^2 + 24x + 41$. $y = 4x^2 + 24x + 41 \neq y = 4x^2 + 24x + 31$ No, both students can't be right. After I expanded the first equation, I come up with 41 as a constant as compared to 31 in the second. I believe the parabolas would have the same shape, etc., but they would be different places on the axes.

Figure 8. Sample item with arguments for students to evaluate

It is often helpful to begin such a task by having students think individually about the problem. Then, as students work to evaluate the responses, they have opportunities to consider different ways the item might be approached. By asking students to make suggestions to improve a response, teachers gain insight into those aspects of the argument on which students are focusing.

In their analysis of exercises related to proof-related reasoning in algebra and precalculus textbooks, Thompson, Senk, and Johnson (in press) found limited opportunities for students to evaluate arguments. Among the 528 exercises out of 9,742 that involved some explicit aspect of proof-related reasoning, only 6 required students to evaluate arguments. So, despite the potential benefit to students of reading and evaluating arguments, students often have limited opportunities with this important task unless teachers make a concerted effort to include such tasks in their instruction.

3.6 *Correcting mistakes in reasoning*

When students evaluate arguments, they do not know whether the argument is valid. So, a variant to evaluating an argument is to tell students that there is an error and direct students to correct the error. Typical misconceptions provide the basis for generating sample student work to be corrected, either as the teacher uses samples of anonymous student work or creates potential responses based on years of experience.

Consider the items in Figure 9. Both sample solutions highlight typical errors that students make (i.e., not reversing the sense of the inequality sign when dividing by a negative, or finding the middle data value without first placing the values in numerical order). Rather than simply asking students to solve the inequality or find the median, the items focus students' attention on finding an error. Because students know they are looking for an error, the tasks are not as open as those in Figure 8 where they needed to evaluate an argument. Nevertheless, students can benefit by thinking about a problem from this different perspective. In the textbook analysis by Thompson, Senk, and Johnson (in press), there were about four times as many exercises for which students were asked to correct a mistake as those in which they were asked to evaluate an argument. Still, the overall percentage of such exercises was small, despite the fact that such tasks provide good opportunities for students to engage with reasoning.

Item A:	Solve: $4x - 7x < -24$.
Modification:	Correct the mistake in the following solution:
	▪ $4x - 7x < -24$
	▪ $-3x < -24$
	▪ $x < 8$
Item B:	Find the median of the data set: 14, 5, 16, 20, 7.
Modification:	Correct the mistake in the following solution:
	I need to find the median of 14, 5, 16, 20, 7. Look for the middle number, which is 16. So 16 is the median.

Figure 9. Modifying an item to correct a mistake

4 Including Reasoning in Assessment

The six techniques in the previous section highlight different strategies that teachers can use to engage students with reasoning, particularly in algebra or statistics concepts. If reasoning becomes a more integral part of classroom instruction, then it also needs to be part of assessment on a regular basis.

The problems in a textbook often serve as the basis of opportunities for students to practice mathematics concepts. The six techniques can be rotated and regularly used to modify one or two questions on typical homework assignments. Student responses to these formative assessments can guide teachers in their instruction. One or two such reasoning tasks can also be added to summative assessments at the end of a unit to emphasize that explaining and justifying one's work is a natural part of mathematics.

5 Conclusion

Reasoning and justification are essential components of mathematics, despite the fact that students often struggle with these processes. As teachers, we must find ways to make these critical processes more prominent. As noted by the National Council of Teachers of Mathematics (2009):

> Reasoning and sense making should occur in every classroom every day. In such an environment, teachers and students ask and answer such questions as "What's going on here?" and "Why do you think that?" Addressing reasoning and sense making does not need to be an extra burden for teachers struggling with students who are having a difficult time just learning procedures. On the contrary, the structure that reasoning brings forms a vital support for understanding and continued learning. ... With purposeful attention and planning, teachers can hold all students in every high school mathematics classroom accountable for personally engaging in reasoning and sense making, and thus lead students to experience reasoning for themselves rather than merely observe it (pp. 5-6).

The six techniques for adapting and modifying items described throughout this chapter are one means of enhancing students' opportunities to engage in reasoning. They are by no means the only techniques (see Kaur & Yeap (2009) for additional techniques). As teachers, our goal should be to ensure that students have many opportunities to focus on reasoning. If curriculum materials provide only limited opportunities for such reasoning, then teachers can use the techniques shared in this chapter to adapt items to make reasoning more prominent. When students make their mathematical thinking visible through reasoning and explanations, teachers gain insight into the robustness of their understanding. Those insights can then be used to modify instruction so that learning is enhanced.

References

Begle, E. (1973). Lessons learned from SMSG. *Mathematics Teacher, 66*, 207-214.

Epp. S. S. (1998). A unified framework for proof and disproof. *Mathematics Teacher, 91*, 708-713.

Hanna, G. (2000). A critical examination of three factors in the decline of proof. *Interchange, 31*(1), 21-33.

Harel, G., & Sowder, L. (2007). Toward comprehensive perspectives on the learning and teaching of proof. In F. K. Lester, Jr. (Ed.), *Second handbook on mathematics teaching and learning* (pp. 805-842). Charlotte, NC: Information Age Publishing.

Herbst, P. (2002). Establishing a custom of proving in American school geometry: Evolution of the two-column proof in the early twentieth century. *Educational Studies in Mathematics, 49*, 283-312.

Herbst, P., & Brach, C. (2006). Proving and doing proofs in high school geometry classes: What is it that is going on for students? *Cognition and Instruction, 24*, 73-122.

Hersh, R. (2009). What I would like my students to already know about proof. In D. A. Stylianou, M. L. Blanton, & E. J. Knuth (Eds.), *Teaching and learning proof across the grades: A K-16 perspective* (pp. 17-20). New York: Routledge.

Johnson, G., Thompson, D. R., & Senk, S. L. (2010). Proof-related reasoning in high school textbooks. *Mathematics Teacher, 103*, 410-417.

Kaur, B., & Yeap, B. H. (2009). *Pathways to reasoning and communication in the secondary school mathematics classroom: A resource for teachers by teachers.* Singapore: National Institute of Education.

Ministry of Education. (2006). *Mathematics syllabus — Secondary*. Singapore: Author.

Movshovitz-Hadar, N. (1998). Stimulating presentations of theorems followed by responsive proofs. *For the Learning of Mathematics, 8*(2), 12-19, 30.

National Council of Teachers of Mathematics. (2009). *Focus on high school mathematics: Reasoning and sense making*. Reston, VA: Author.

National Council of Teachers of Mathematics. (2000). *Principles and standards for school mathematics*. Reston, VA: Author.

Sinclair, N. (2008). *The history of the geometry curriculum in the United States.* Charlotte, NC: Information Age Publishing.

Stylianou, D. A., Blanton, M. L., & Knuth, E. J. (Eds.). (2009). *Teaching and learning proof across the grades: A K-16 perspective*. New York: Routledge.

Thompson, D. R., Beneteau, C., Kersaint, G., & Bleiler, S. K. (in press). Voices of mathematicians and mathematics teacher educators co-teaching a mathematics course for prospective secondary teachers. In J. Bay-Williams & W. Speer (Eds.),

Professional collaborations in mathematics teaching and learning. Reston, VA: National Council of Teachers of Mathematics.

Thompson, D. R., & Senk, S. L. (1993). Assessing reasoning and proof in high school. In N. L. Webb & A. F. Coxford (Eds.), *Assessment in the mathematics classroom* (pp. 167-176). Reston, VA: National Council of Teachers of Mathematics.

Thompson, D. R., Senk, S. L., & Johnson, G.J. (in press). Opportunities to learn reasoning and proof in high school mathematics textbooks. *Journal for Research in Mathematics Education*.

Valverde, G. A., Bianchi, L. J., Wolfe, R. G., Schmidt, W. H., & Houang, R. T. (2002). *According to the book: Using TIMSS to investigate the translation of policy into practice through the world of textbooks*. Dordrecht, The Netherlands: Kluwer Academic Publishers.

Yackel, E., & Hanna, G. (2003). Reasoning and proof. In J. Kilpatrick, W. G. Martin, & D. Schifter (Eds.), *A research companion to Principles and Standards for School Mathematics* (pp. 227-236). Reston, VA: National Council of Teachers of Mathematics.

Chapter 6

LOGO Project-Based Mathematics Learning for Communication, Reasoning and Connection

Hee-Chan LEW In-Ok JANG

This paper introduces the project-based mathematics learning through LOGO programming activities, which was designed for Korean 6th grade (Year 6) promising students to improve their mathematical reasoning skills and to activate communication with peer students and teachers in various projects to connect mathematics and visual art. In this learning process, students were found to activate and promote reasoning strategies such as analogy, generalization, progressive and critical thinking and debugging based on visualization and empirical inference. Students could activate their communication with other students and their teacher in analyzing, debugging, comparing and contrasting their programming. This chapter shows that supporting with the peculiar characteristics of LOGO language, LOGO project-based learning is an effective environment to combine Piaget's epistemology on intelligence development, Polya's heuristics for improving mathematical problem solving abilities and Poincare's philosophy to emphasize mathematics education to nurture attitudes towards mathematics, intuition and esthetic sense.

1 Introduction

LOGO, which was developed by S. Papert of MIT in 1980, can be explained by two metaphors for effective mathematics learning (Papert,

1980). One is the metaphor of "math-land," denoting a good place to understand abstract mathematical concepts like angle, length, variables, and functions through concrete activities and to a matrix of further higher abstract concepts. The other is "mind-storm," referring to a tool to provoke and guide students' thinking in the micro-world. To actualize these metaphors LOGO engages four characteristics as programming language, namely body-syntonic, procedural, mathematical and recursive. With these characteristics, LOGO provides an easy and natural environment that enables students to reflect on their own thinking and eventually encourages project-based mathematics learning.

After describing these four characteristics, this chapter will introduce some results of LOGO project-based mathematics learning, which was designed for Korean 6th grade (Year 6) students to improve their mathematical reasoning skills and to activate communication with peer students and teachers in various projects to connect mathematics and visual art. In addition, this chapter will show that LOGO project-based mathematics learning is an effective way to combine Piaget's epistemology on intellectual development as a continuous process of reflective abstraction, Polya's heuristics as thinking skills for improving problem solving abilities and Poincare's philosophy that emphasizes the role of mathematics education in nurturing students' attitudes towards mathematics, intuition and esthetic sense.

2 Characteristics of LOGO

For a long time, prominent efforts to improve mathematics learning by using computers in the classroom have involved LOGO activities (Swan, 1989; Keller, 1990; Kim, 1992; Subhi, 1999; Shin, 2000; Clements, Sarama, Yelland, & Glass, 2008). These roles of LOGO in mathematics education come from the following distinctive characteristics.

First, LOGO is closely related to students' actions: body-syntonic. The movement of a turtle implemented by a LOGO command is matched easily to students' actions in the thinking level. For example, when the command below is given, even students who are less experienced with computers can easily imagine that the result is a square with the length of

50. And, conversely, when the square with a length of 50 is given, students can easily generate the command for a square.

```
REPEAT 4 [FD 50 RT 90]
```

This characteristic enables students to embody abstract geometrical concepts like squares, distances, angles. The embodiment reached through LOGO programming activities work as an operational tool for helping students induce understandings of abstract concepts in the higher level because such understandings are accomplished consciously on a concrete level. LOGO makes not only the joining of abstraction and concreteness for a higher level of thinking possible but can also reduce the sense of difficulty when they learn about mathematics on an abstract level.

Second, LOGO is a "procedural" language. Once a procedure to draw a certain figure is written, it is stored as one "name" to be used like basic commands in other future programming tasks. For example, when the commands to draw an equilateral triangle and to draw a square are named as "TRI" and "SQ" respectively, TRI and SQ are used like a basic command in the procedure to draw the figure of a house (*Figure* 1).

```
TO TRI
REPEAT 3 [FD 100 RT 120]
END
TO SQ
REPEAT 4 [FD 100 RT 90]
END
TO HOUSE
SQ FD 100 RT 30 TRI LT 30 BK 100
END
```

Figure 1. Procedure to draw a figure of house

This procedural characteristic makes students aware of and provokes them to use their previous thinking or actions in the programming process. Such awareness and use of previous thinking or actions are psychologically and mathematically very important in the sense that this is one kind of abstraction and at the same time an operational process for higher abstraction. This is also educationally very important in that an "operation of abstraction" is one methodology that can reduce the sense of difficulty students feel.

This characteristic makes "structured" programming possible in LOGO. A whole programming process utilized for a complex problem task can be divided into several functional units which are analyzed and implemented independently and then connected later. Eventually, the structured language makes the programming and debugging process much easier and simpler.

Third, LOGO is a mathematical language, which means that in the programming process, the shapes turtle draws are determined by the values of the variable. For example, if we put a variable of the length of sides in the procedure to draw an equilateral triangle, students can draw triangles with a specific length according to the values of the variable (Figure 2). In this process students encapsulate various numbers in one variable and recognize that variable is not just a letter but a special symbol to contain many elements which have a common meaning.

```
TO TRIANGLE : X
REPEAT 3 [FD : X RT 120]
END
```

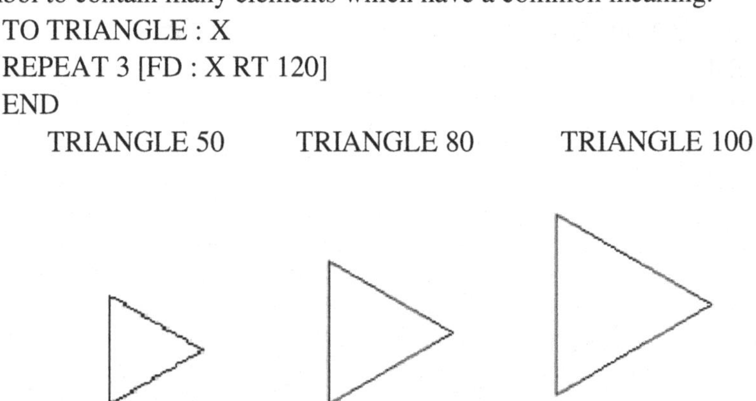

Figure 2. Triangles with length of sides as the variable

Here, more than two variables can be used and unlike variables that have traditionally appeared in mathematics textbooks, ordinary names can be used in the procedure. For example, in the following procedure, two variables are used and variable names are usual language. Here, "side" means the number of polygon's sides and "multi" represents the points of a polygon to be connected (Figure 3). This characteristic is important in mathematics education as it alludes to how the concept can be taught meaningfully and effectively. The concept of a variable is supposed to be difficult for students to learn. The concept of functions is also said to be difficult for ordinary students. LOGO provides a very effective learning environment that helps reduce the difficulties students experience in learning about such concepts as variables and functions.

```
TO POLY : SIDE : MULTI
REPEAT : SIDE [FD 100 RT 360 * : MULTI / : SIDE]
END
```

POLY 7 2 POLY 10 3 POLY 12 5

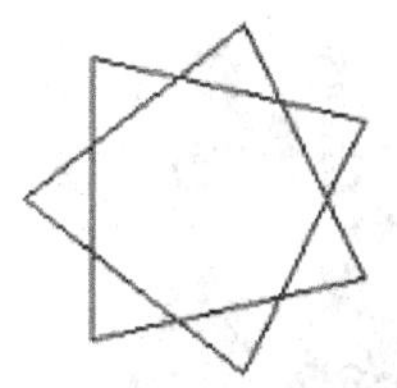

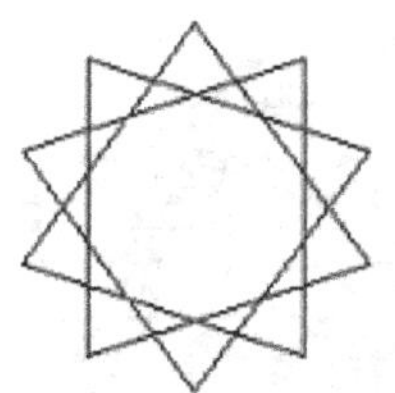

Figure 3. Procedure with two variables

Finally, LOGO is a "recursive" language. That is, one procedure can be used as a command in itself. Recursion sets up a never-ending process. For example, in the following procedure named as "poly1," the "poly1" itself is used once more and the recursive procedure triggers a continuous repetition of [FD 100 RT 72], and the final result is a pentagon (Figure 4).

As Papert (1980) claims, a recursion is particularly able to evoke an excited response because the idea of "going on forever" touches on every child's fantasy. It provides opportunities to deal with the concept of infinity sensuously and make children feel like mathematicians. And, this characteristic of LOGO can make beautiful figures that are beyond general expectations (Figure 5).

```
TO POLY1
FD 100 RT 72  POLY1
End
```

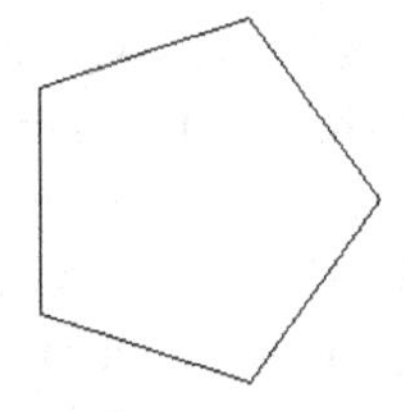

Figure 4. Recursive procedure of pentagon

```
TO STAR : X : R : N
IF : N = 0 [STOP]  LT 126REPEAT 5 [FD : X LT 18 STAR : X * :
R : R : N - 1 RT 18 RT 144 FD : X LT 72]  RT 126
END
 STAR 30 0.4 3          STAR 30 0.4  4          STAR 30 0.4  5
```

Figure 5. Recursive procedure of STAR

What needs to be noted here is the role of LOGO in inducing students to reflect on their own thinking. With the four characteristics mentioned above, LOGO provides an easy and natural environment that encourages students to be aware of their actions or thinking, to analyze or debug them, and to generalize or encapsulate actions performed previously. This is because the educational mechanism that supports LOGO was based on the reflective abstraction clarified by Piaget (Blackwelder, 1985). According to Piaget (1973), intellectual development is a continuous process of reflective abstraction which is drawn from the coordination of several actions including awareness, analysis and generalization unlike the physical abstraction which derives its sources from the physical properties of objects. Papert believes that LOGO is a math-land in which children can learn mathematics naturally and spontaneously based on reflecting on their own thinking, just like the environment in which children learn their mother tongue.

Furthermore, programming in LOGO is based on the Polya-style problem solving process. Undergoing steps of understanding, planning, carrying out, and looking back, students are exposed to the problem solving process (Polya, 1957). Problem solving strategies, like to subdivide and to relate to already known facts, can be exercised in a natural setting (Papert, 1980). Particularly, students can foster their meta-cognitive skills or managerial skills for problem solving by reflecting on their own thinking or debugging.

A by-product of LOGO is that students' intuitive thinking can be improved by grasping the situations in their own eyes. Reflecting on their own thinking, students develop a sound mathematical attitude, their own mathematical intuition and mathematical esthetic sense, which are important objects of mathematics education according to the French mathematician H. Poincare (1963).

Papert (1980, 1991) proposes constructionism related to the growth of knowledge based on the epistemological positions by Piaget, Polya and Poincare. Its fundamental concept is an assimilation process of new knowledge to the existing scheme and the process is established through an interaction between subject and environment rather than through didactic teaching. The interaction is facilitated by a self generating process through learning by doing, communication among teachers and students, inductive and deductive reasoning and integration of thinking. He argues that LOGO is an interactive environment of which role cannot be easily played by the existing school system. Furthermore, the connection between mathematics and other subjects like art, or between geometry and algebra can be done naturally because LOGO is basically based on drawing by turtle and control by variables.

In this sense, LOGO is a good environment for pursuing the educational objectives that have been recently emphasized in Korea's mathematics curriculum. According to the mathematics education policy issued by the Korean MEST (2011), the connection between mathematics and other subjects like science, technology, engineering, and art which are together referred to as STEAM is to be accentuated so that students and the general public can have a positive outlook towards mathematics. It is suggested that this can be accomplished by developing a more interesting mathematics education that is connected to everyday life. In addition, integrated thinking or reasoning, various problem solving activities and reasonable communication in the mathematics classrooms are emphasized to train creative manpower and to construct a democratic society. That is, the main point of the Korean government's mathematics education strategy is to strengthen communication, reasoning and connections in the mathematics classroom. Many scholars also indicate that the education process has to place importance on diversity in depth and range for instructional content to provoke students' intellectual curiosity and promote integrated thinking to connect knowledge of diverse fields (Samara, Pedraza & Curry, 1992; Sheffield, 1994; Make & Nielson, 1995; Van Tassel-Baska, 1994).

3 LOGO Project-Based Learning for the Elementary Students

The remaining pages introduce LOGO project-based mathematics learning as a good methodology for fostering skills in communication, reasoning and connections shown by some results applied to the 6th grade Korean students at the Education Center for the Scientifically Gifted at the Seoul National University of Education (Jang, 2011). They are "promising" students as they make up the top 5% in their achievement scores in their schools. However, the learning process can be applied similarly to ordinary students by reducing the difficulty level of contents used in the classroom and by providing an enriching treatment to more actively engage when students are confronted with problems. Four students were chosen from a group of 40 students by teacher recommendation and student interviews. These four formed two groups of two students by their familiarity.

LOGO project based learning consists of a series of unit learning in which students independently draw diagrams required in completing the works of a selected theme given as project tasks. Through unit learning, students communicate with each other to search for the most efficient method to draw a diagram and to find various solution methods with more interesting ideas. The process to set forth the project theme and a plan for programming to draw the appropriate diagrams would basically be the connection of mathematics and art and would require various kinds of reasoning to design or modify the procedure for several diagrams in the project. In addition, the process is closely related to such higher level thinking activities as critical and analytic thinking.

Many previous studies also consider that LOGO can be used in an effective learning environment to improve higher levels of thinking (Clements & Gullo, 1984; Clements et al., 2008; Keller, 1990; Swan, 1989). However, this study will focus on what strategic thinking the mathematically promising 6th grade students used to plan, implement and debug in the programming process as a problem solving process. The students learned the MSWLOGO commanding language to process the recursive procedure and to execute an animation by using variables and conditional phases to complete the project during a total of 12 experimental classes of 2 hours each, as shown in Table 1.

Table 1

Whole schedule of LOGO project-based learning

Whole schedule of LOGO project-based learning
Theme setting: Class [1]-[3]
The students first learned the basic commanding language and defined the procedures to draw several diagrams by using commanding language in the first 2 classes, which then informed the project learning in the third class. Students cooperated to determine the project theme for each group, what diagrams to draw and how to draw them. [See the Appendix 1 for the classroom activities of the classes [1]- [3]]
Diagram structuring: Class [4]-[9]
In the 4th to 9th classes students built up programming of diagrams provided by the teacher and accomplished various works selected by themselves or tasks to modify diagrams provided by teachers for the project for each group. [See Appendix A for the classroom activities of the classes [4]- [9]]
Project completion: Class [10]-[12]
In the 10th to 12th classes, students devised activities to integrate several diagrams in each group. The previously made diagrams were modified or new diagrams were made if necessary. [Final results of two groups are shown in Appendix B.]

4 Some Results of the Pilot Lesson Study

In this pilot lesson study, it was verified that LOGO is a good environment for students to encourage their communication and reasoning skills. LOGO is incorporated in dynamic project-based learning that provides students with opportunities to apply and develop their mathematical knowledge and engage in diverse creative activities through the integration of mathematics and art as a positive way to foster higher levels of thinking to plan, implement and debug.

4.1 *Planning and implementing strategies*

Analogy

Students were asked how to draw the two diagrams shown in Figure 6. Students who developed the recursive procedure to draw the maze (Figure 7) in the previous time in the 6th class thought the diagram was similar with the maze procedure in that the length of the side is changed but was different than the size of the internal angle of a regular polygon.

Based on the similarity and the difference, they could draw the above two diagrams presented by the teacher. They understood that the angle size for drawing the two diagrams is related to the internal size of the regular square and were also able to conclude that the left diagram has a rotation angle smaller than 90 degrees (POLYGON 5 89 2) and the right diagram has the size of rotation angle larger than 90 degrees (POLYGON 5 91 2). Students could engage in analogical thinking based on the procedure to draw the maze.

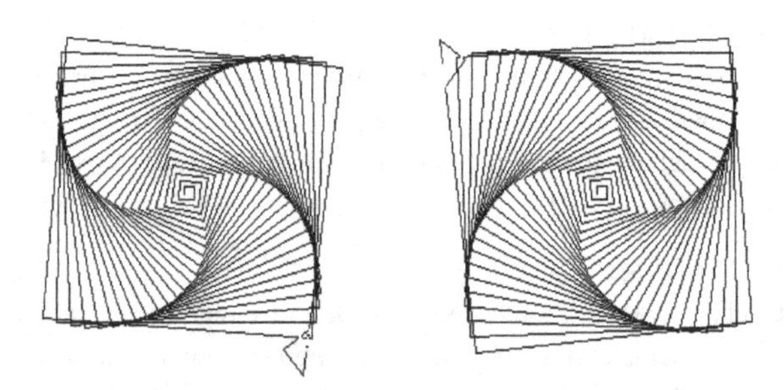

Figure 6. The two diagrams students were asked to draw

```
TO POLYGON : S : A : I
IF : S>200[STOP] FD : S RT : A  POLYGON : S + : I : A : I
END
                    POLYGON 5 90 2
```

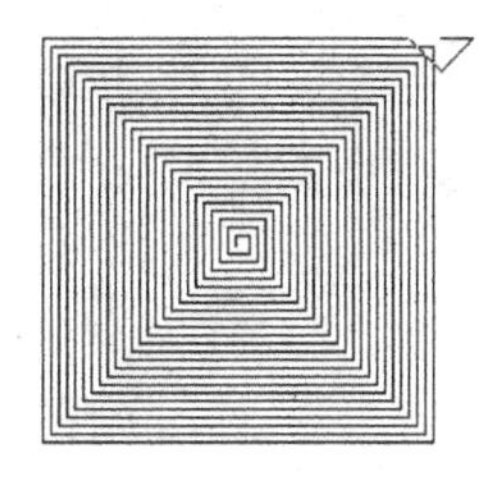

Figure 7. Recursive procedure of maze

Generalization

In the 5th class to draw various figures using circles and several regular polygons, Soohyun drew the vehicle to use in the project (Figure 8). After drawing a special sized vehicle, she determined the method to generalize the procedure for drawing the vehicle. The special sized vehicle seemed to be difficult to generalize as the circle and polygon were mixed, but students understood that the generalization of the procedure is convenient to draw several shapes and they had a clear grasp of the meaning of generalization as a formula in mathematics.

Soohyun proposed to input a variable in the vehicle procedure while controlling the size. Eventually, they decided to change each number to represent distance to the numbers divided by 10 multiplied: X. They did not change the number to represent angles (Figure 8). They made a procedure as a general formula.

Critical thinking

After learning how to draw a circle by the procedure REPEAT 360 [FD 1 RT 1] in the 2nd class, the students were able to draw different types of shapes by using the circle. However, Juseong pointed out in the 5th class that the figure which was a mixture of a regular polygon and circle was difficult to draw because the length of the radius was inaccurate. He determined that in order to draw a mixed diagram of a circle and straight line, for example a semi-circle, the length of the radius had to be known.

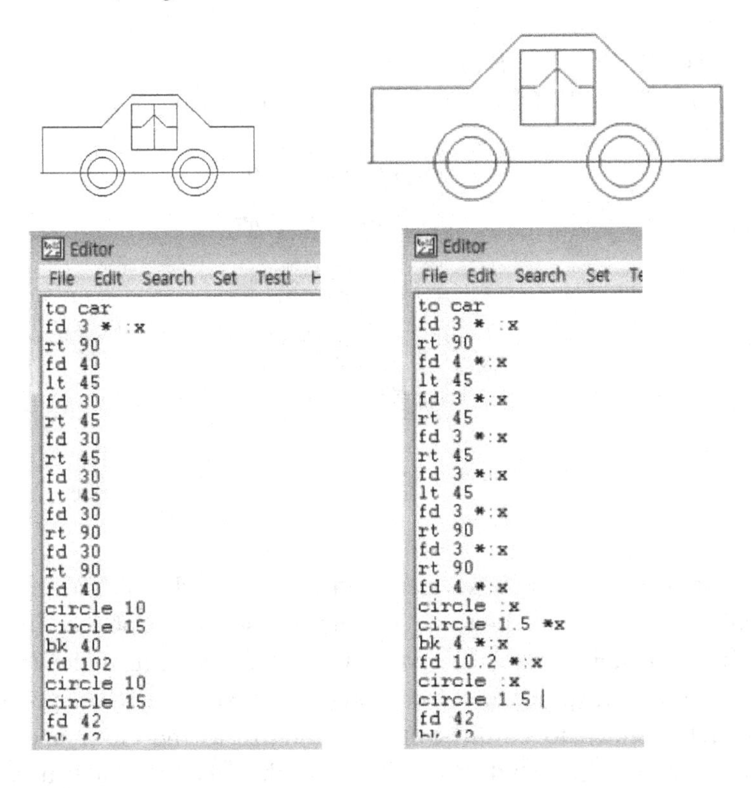

Figure 8. Generalization process for various vehicles

Such analysis motivated the students to investigate the procedure to draw a circle using the radius, which means that their role had changed to that of a designer making new programming language using their critical thinking skills. Juseong proposed the new procedure using the below figure (Figures 9, 10).

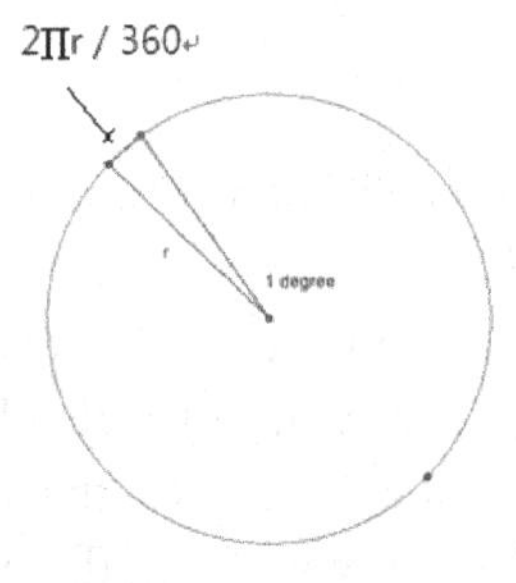

Figure 9. The length of arc for the central angle of 1 degree

```
TO CIRCLE : r
PUFD : rRT 90 PDREPEAT 360 [FD 2*3.14* : r/360 RT 1] LT 90
PUBK : r PD
End
```

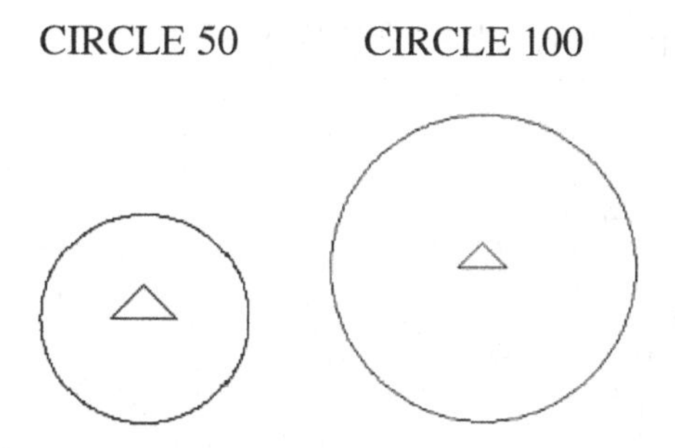

Figure 10. Recursive procedure of circle

Progressive thinking

Juseong intended to draw a stair-shaped tree using the regular triangle for the project and tried to establish an appropriate procedure in the 4th class. However, because the directions to draw the first triangle and the second triangle differed the desired diagram could not be drawn (Figure 11). Rather than being disappointed the students were able to grasp this as an opportunity to make a new shape.

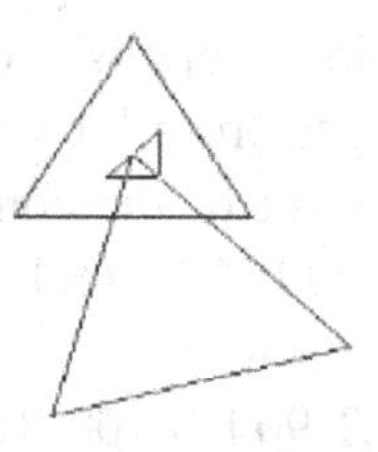

Figure 11. Unexpected procedure of the stair-shaped tree

With the help of the teacher who asked about how they could draw a golden spiral tree, the students were able to understand that the diagram of the golden spiral shape could be drawn by using the conditional procedure to change the length of the side and the size of the angle. Juseong drew a golden spiral shape in which a triangle rotated and became gradually smaller. Soohyun extended it by using a square instead of a triangle. Interpreting the diagram in error from a different point of view was transformed into an unintended learning opportunity (Figure 12).

To SPIN: x : a
IF: x = 200 [stop] REPEAT 3 [FD: x RT 120] RT: a SPIN: x + 10: a
End

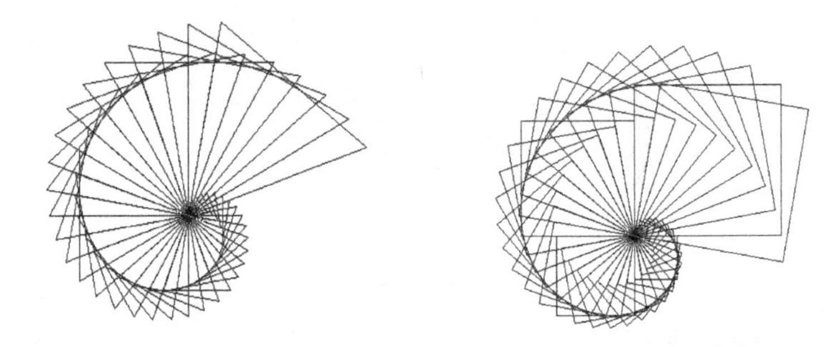

Figure 12. Recursive procedure of golden spirals

4. 2 *Debugging strategies*

Visualization

Jeho drew the windmill rotating toward the right direction by the following command in the 3rd class (Figure 13).When he was asked to draw the windmill rotating in the opposite direction he thought that if the rotation direction was changed from LT to RT in the procedure it could be drawn easily. But, the result of the command REPEAT 12 [FD 100 RT 90 FD 100 BK 100 RT 90 BK 100 RT 30] came out as the sunflower in Figure 14.

REPEAT 12 [FD 100 LT 90 FD 100 BK 100 RT 90 BK 100 RT 30]

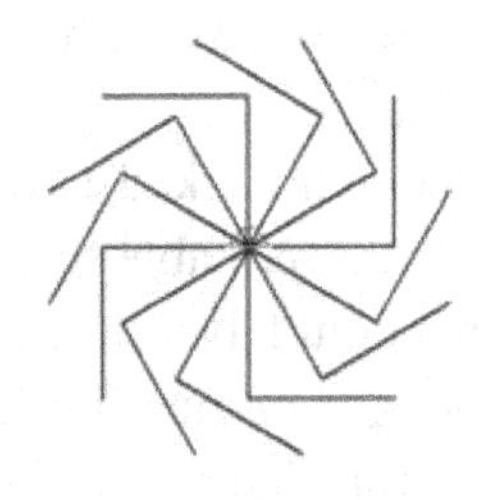

Figure 13. Recursive procedure of windmill rotating towards the right direction

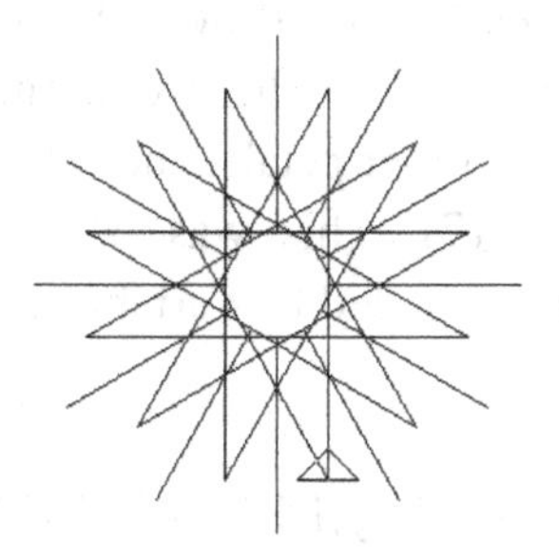

Figure 14. Wrong recursive procedure of windmill rotating towards the left direction

Watching this shape, Juseong found out that the unexpected diagram was connected complicatedly and based on this visualization toward the new task the rotation direction is also changed from RT to LT when turtle returned to the original home position by BK to make the following command (Figure 15). Juseong clearly understood that which part of the turtle changes depended on the direction of the turtle. This is because of the body-syntonic characteristic of LOGO.

REPEAT 12 [FD 100 RT 90 FD 100 BK 100 **LT** 90 BK 100 RT 30]

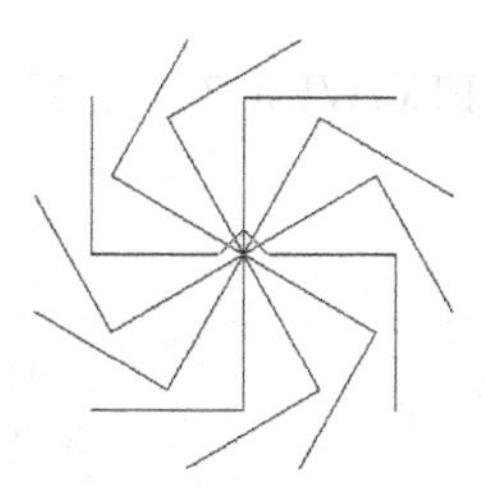

Figure 15. Recursive procedure of windmill rotating towards the left direction

Empirical inference

Soohyun constructed the general procedure to draw a flower with petals provided by the teacher by using circles and arcs in the 5th class (Fig.16). In order to draw a flower on the screen, a specific value (50, 30) of the two variables was inputted on the general procedure, but it did not work. She found that an error occurred by the selected values of variables rather than the procedure itself. Then she proposed an idea by saying that "I have a good idea. Let's execute in sequence from flower 1 1 which is the smallest value." It was selected as a part of the way of finding the proper values for the flower. Watching the shape with the variable value 1 1 that is, based on her experience, she drew an inference of the rotation numbers of other petals, and the radius sizes of the desired shaped flowers (Figure 17). This was possible because LOGO is a mathematical language controlled by variables.

```
TO FLOWER: x : y
CIRCLE: x*23 REPEAT: y[REPEAT 2 [REPEAT 90[FD: x RT 1]
RT90] RT 360/: y]
End
```

Figure 16. Recursive procedure of flower

```
FLOWER 2 12 RT 15 FLOWER 2.5 12 RT 15 FLOWER 3 12
```

Figure 17. Debugging process by empirical inference

5 Conclusion

It is apparent from the pilot lesson study that firstly, project learning using LOGO can be utilized in an integrated curriculum. The work for each theme made with diverse types of graphics can provide a meaningful learning experience that integrates logical elements of mathematics and aesthetic elements of art in completing a final product in each class, therefore it is an overall integrative, cohesive and systematical gifted education program. Thus, it can be applied as an alternative to current educational curricula for the gifted in mathematics in that it allows comprehensive and meaningful learning of the internal value of logo language, as well as fostering an in-depth understanding and applicability of mathematical concepts like properties of polygon, conversion of shape, understanding on angle, the concept of variables.

Secondly, LOGO project learning can be facilitated as an efficient educational program to further improve creative problem solving skills of gifted students. Analogical thinking, generalization, critical thinking, progressive thinking, flexible thinking, visual inference and empirical inference demonstrated in the LOGO programming process represent higher-level thinking capabilities. This implies that learners' experience of programming diagrams in various ways presented by their teacher or independently planned unique diagrams have a close relationship with such thinking activities.

Thirdly, analysis of higher thinking skills displayed in the LOGO project learning can be used as the basis for developing curriculum materials for teaching and learning specially for promising student and also for all students in mathematics. Materials used in this LOGO project learning can used as the instructional model for considering what type of strategic thinking should be emphasized for LOGO learning project and how the learning and teaching process should be accomplished in school mathematics.

References

Blackwelder, C. K. (1985). *LOGO: A possible aid in development of Piagetian formal reasoning*. Unpublished doctoral dissertation, Georgia State University.

Clements, D., & Gullo, D. (1984). Effects of computer programming on young children's cognition. *Journal of Educational psychology, 76*(6), 105-158.

Clements, D., Sarama, J., Yelland, N. J., & Glass, B. (2008). Learning and teaching geometry with computers in the elementary and middle school. In M.K. Heid & G. W. Blum (Eds.), *Research on technology and the teaching and learning of mathematics: Vol. 1. Research syntheses* (p.109-154). Charlotte, NC: Information Age.

Jang, I. O. (2011). The strategic thinking of mathematically gifted elementary students and teacher's role in LOGO project learning. Unpublished Doctoral Dissertation, Korea National University of Education.

Keller, J. K. (1990). Characteristics of LOGO instruction promoting transfer of learning: A research review. *Journal of Research on Computing in Education, 23*(1), 55-71.

Kim, S.H. (1992). An analysis on the effects of LOGO programming for the development of logical thinking. Unpublished Master Thesis, Korea National University of Education.

Korea Ministry of Education, Science and Technology (2011). *A plan for upgrading mathematics education.* Unpublished manuscript.

Make, C., & Nielson, A. B. (1995). *Teaching models in education of the gifted.* Austin, TX: Pro-Ed.

Papert, S. (1980). *Mindstorms: Children, computers, and powerful ideas.* The Harvester Press Limited, Sussex.

Papert, S. (1991). Situating constructionism. In I. Harel & S. Papert. (Eds.) *Constructionism* (pp. 1-12). Norwood. NJ: Ablex.

Piaget, J. (1973). Comments on mathematics education. In A. G. Howson (Ed.), *Development in mathematics education (proceedings of 2nd ICME)* (pp. 79-87). Cambridge University Press.

Poincare, H. (1963). *Mathematics and science last essay.* New York: Dover Publishing.

Polya, G. (1957). *How to solve it.* Princeton: Princeton University Press.

Samara, J., Pedraza, C. & Curry, J. (1992). *Designing effective middle school units.* Glenview. IL: Association for Gifted Children.

Sheffield, L. J. (1994). *The development of gifted talented mathematics students and the NCTM Standards.* Storrs, CT: The National Research Center on Gifted and Talented, University of Connecticut.

Shin, H. J. (2000). A case study on understanding of variable concept for the 6th graders through LOGO programming. Unpublished Master Thesis, Korea National University of Education, Seoul, South Korea.

Swan, K. (1989). LOGO programming and the teaching and learning of problem solving. Unpublished doctoral dissertation, Teachers College, Columbia University.

Subhi, T. (1999). The impact of LOGO on gifted children's achievement and creativity. *Journal of Computer Assisted Learning, 15,* 98-108.

VanTassel-Baska, J. (1994). *Comprehensive curriculum for gifted learners.* Needham Height, MA: Allyn & Bacon.

Appendix A: Tasks Used in LOGO Project-Based Learning

1st class

- Introduction to MSWLOGO and 3 kinds of screen
- Basic commands like FD, BK, RT, LT, PU, PD, PE, HT, ST, Home

2nd class

- A regular triangle, a regular pentagon,a circle
- Double repeats, for example Repeat 18 [Repeat 360 [FD 1 RT 1] RT 20],Procedure like Square, Triangle, House, Repeat of House
- Procedure with variable(s) like Regular n-polygon, for example, Polygon :n, Polygon :x :n
- Command for coloring blocks

3rd class

- Application of Square Procedure(Basic)
- Parallelogram, Rectangle
- Application of Square procedure(Advanced)
- Make Command: Make "Name X
- Label Command: Label "Word or sentence

4th class

- Application of Triangle Procedure(Basic)
- Application of Triangle Procedure(Advanced)

5th class

- Application of Circle Procedure(Basic) like Translation of circle, Circles with a same center,
- Application of Circle Procedure(Advanced) like Flower, Petal

6th class

- Recursive Procedure(Basic) like Spider-web , Maze
- Recursive Procedure(Advanced) like Rotation of Square Maze, Rotation of Flowers, Rotation of Polygons

7th class

- Recursive Procedure(Advanced 2)
- Fractal like Koch curves, Snow flake, Sirpinski Triangle, Tree

9th class

- Animation of car and wheel

Appendix B: One Final work of LOGO Project-Based Mathematics Learning

Chapter 7

Reasoning, Communication and Connections in A-Level Mathematics

TOH Tin Lam

After the curriculum review in 2006, the Singapore school mathematics framework expanded its scope to include reasoning, communication and connections as three main components of the **Processes** of problem solving. This new emphasis should pervade all levels of mathematics learning: primary, O-Levels and A-Levels. As A-Level students have learnt sufficient elementary mathematics at the O-Level and are preparing for tertiary education, this emphasis on reasoning, communication and connections is not less important than at the O-levels. In this chapter, sample activities of how reasoning, communication and connections can be infused into teaching various A-level mathematics topics are discussed. For mathematical reasoning at the A-levels, elementary mathematical proofs and derivation of results by first principles are good opportunities to transcend procedural emphasis of traditional teaching to higher level mathematical reasoning. Connections can be achieved by several means: facilitating students to connect across different mathematical ideas, across other disciplines and to daily life. Connections help students to better understand and appreciate various mathematical concepts and enable them to regulate their own thinking (meta-cognition). Teachers should provide opportunity for students to communicate their reasoning clearly, and formulate their argument during mathematics lessons.

1 Introduction

The well-known pentagon framework of the Singapore school mathematics curriculum, which places problem solving as its heart, has been in existence since the 1980s. The recent refinement of the school mathematics framework in 2007 (Ministry of Education, 2006) has expanded the scope of **Processes** to include mathematical reasoning, communication and connections as three key components (Figure 1).

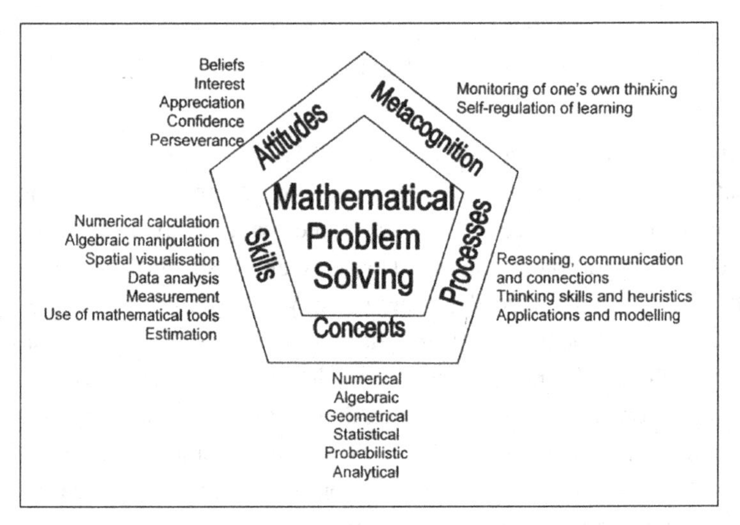

Figure 1. Framework of the school mathematics curriculum

According to the Mathematics syllabus document, issued by the Ministry of Education (MOE), "Mathematical reasoning, communication and connections should pervade all levels of mathematics learning, from the primary to A-levels" (MOE, 2006, p. 4). This shows that the latest expansion in scope for the processes to include mathematical reasoning, communication and connections is applicable to A- levels.

Students entering A-level institutions have learnt significant mathematics at the secondary levels and are preparing for their tertiary education. Thus, the emphasis on mathematical reasoning, communication and connections must be taken seriously by teachers in planning their lessons.

Unlike primary and secondary schools, mathematics teachers' guidebooks aligned to the curriculum and textbooks to heighten teachers' awareness of the new emphasis of the latest syllabus are not easily available for A-level mathematics teachers. Thus, A-level mathematics teachers might face the challenge of designing their teaching material and mathematical tasks that are in line with the latest emphasis of the curriculum.

In this chapter, feasible teaching ideas and sample mathematical tasks to incorporate reasoning, communication and connections in the teaching of A-level mathematics are introduced.

2 Mathematical Reasoning

Mathematics should make sense to students, and provide students with ample opportunity to seek and find explanations for their observations. According to the *Principles and Standards for School Mathematics* published by the *National Council of Teachers of Mathematics* (NCTM) reasoning is: "...not a special activity reserved for special times or special topics in the curriculum but should be a natural, ongoing part of classroom discussions, no matter what topic is being studied" (NCTM, 2000, p. 342).

Through reasoning, students could develop an appreciation of mathematical justification in the study of the mathematical content. The higher the educational level of the students, the more sophisticated should the methods of mathematical reasoning involve. Two different ways to infuse reasoning into teaching A-level mathematics are discussed in this section: refining the way to teach procedural tasks, and engaging students in more in-depth understanding of mathematical proofs (taught at the A-levels).

2.1 *Mathematical reasoning of procedural tasks*

Making logical arguments and analysing mathematical situations should be the core part of teaching A-level mathematics, even in the most procedural portion of the mathematics content. As an example, consider the topic "Differentiation Techniques", in which most differentiation formulae require students to commit to memory. A commonly used formula is

$$\frac{d}{dx}(\sin x) = \cos x,$$

with the condition that x must be measured in radians.

A survey of the lecture notes and tutorial worksheets designed by teachers reveals that few teachers would want to delve deep into the formula through the first principle for this specific formula, or engage their students in exploring why the unit of measurement in radian is crucial for this formula to be true. Often students are expected to memorize this formula, including the condition that the angle in this differentiation formula must be measured in radians.

A study carried out by Toh (2009) shows that when 27 practicing secondary mathematics teachers were asked to obtain a formula for the derivative of sin x°, 17 teachers gave the answer as

$$\frac{d}{dx}(\sin x^{\circ}) = \cos x^{\circ}, \qquad (1)$$

which clearly demonstrates that the practicing teachers had forgotten the condition about the unit of measurement of the angle. Even practicing teachers have difficulty memorizing the condition appropriate to apply the formula, what more about A-level students!

How did the participants generalize the formula in (1)? When one encounters a new problem, a mental structure which contains *tentative solution starts* is created (Selden, Selden, Hauk & Mason, 1999). Without further inputs through sound mathematical reasoning, learners are likely to form their own "generalization" not based on sound reasoning, but on visual semblance. This explains how the "generalized result" in (1) is formulated based on the learners' own interpretation of the derivative of the function sin x°.

On the other hand, one could have a better appreciation of the derivative of sin x in (1) if one has gone through the derivation by the first principle, and paid attention to a critical step involving the use of the identity

$$\lim_{x \to 0} \frac{\sin x}{x} = 1$$

which is true only if x is measured in radian. A rigorous derivation of this limit is not required (or encouraged) at the A-levels, but an appreciation of this limit could help students to recall the concepts related to this limit better. A sample of the reasoning that can accompany the derivation of the formula by the first principle is shown in Figure 2.

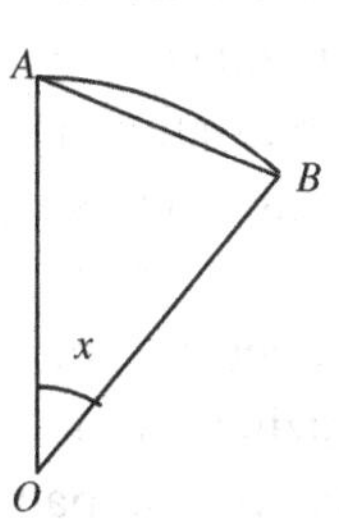

Consider the sector OAB which subtends an angle of x radian at its centre O.

a. Find expressions for the area of the triangle OAB and the sector OAB.

b. What can you tell about the areas of triangle OAB and the sector of OAB when x is small?

c. Can you tell the value of $\lim_{x \to 0} \frac{\sin x}{x}$?

d. Suppose now that x is measured in degrees. Find similar expressions for the area of triangle OAB and the sector OAB.

e. If the angle x is small, what can you tell about the areas of triangle OAB and the sector OAB when x(in degrees) is small?

f. In this case, can you tell me the value of $\lim_{x \to 0} \frac{\sin x^{0}}{x}$?

g. Based on the above and the first principle of differentiation, what do you predict is the formula for $\frac{d}{dx} \sin x^{0}$? Go back to the first principle and demonstrate how you got your formula.

Figure 2. Sample of a mathematical reasoning task related to deriving the derivative of the sine function

The above task could be offered as a hands-on or self-discovery activity prior to the students learning differentiation of trigonometric functions. Engaging students in this task could serve to activate their prior knowledge (students have learnt areas of sectors in O-Level mathematics) and subsequently acquire an in-depth appreciation of the formula and its derivation. One who has an in-depth relational understanding of mathematical results would more likely be able to remember and appreciate the formula, compared with one who has mere procedural knowledge of the formula (Skemp, 1976).

2.2 *Reasoning and mathematical proofs*

The role of a mathematics teacher in helping students to develop reasoning includes "...convey the importance of knowing the reasons for mathematical patterns and truths... to evaluate the validity of proposed explanations..." (NCTM, 2000, pp. 345-346). Different levels of mathematical reasoning are emphasized at different levels of the mathematics curriculum: at the secondary level, making conjectures and looking for patterns, form the bulk of the mathematical reasoning; at the A-levels, mathematical reasoning consists of understanding some simple formal mathematical proofs.

From a mathematician's perspective, making conjecture about a formula based on inductive reasoning or pattern gazing is not an acceptable mathematical "proof": a well-known example that pattern gazing is not "proof", and may sometimes mislead, can be found in Toh, Quek, Leong, Dindyal & Tay (2011, p. 125, problem 15). It is crucial that such an example of "pattern gazing is not proof" be impressed upon A-level students in preparing them to accept formal mathematical proof.

On the other hand, the informal proof schemes at the secondary level should not simply be dismissed as incorrect, but rather be accepted as facets of mathematical reasoning necessary to achieve and master mathematical argumentative practices (Recio & Godino, 2001). A-level teachers should build on these informal proof schemes to introduce higher level rigorous mathematical reasoning. We next illustrate how this could be done.

Students first encounter formal mathematical proofs in the A-levels. The progression from informal to formal proof is a 'natural' progression for students learning mathematical proofs and reasoning. The lack of opportunity for students to appreciate this natural progression could possibly lead students to have two different conceptions of proof: those about arguments they consider would be awarded the best score in examinations and those which they would adopt for themselves (Healy & Hoyles, 2000).

In order to help students achieve a smooth transition advancing from informal to formal proof, teachers could modify the existing standard examination-type proof questions to make them more "natural" to the students through mathematical investigation and exploration. Consider Questions A and B in Figure 3.

Question A: A sequence $u_1, u_2, \ldots, u_n, \ldots$ is defined by the formula $u_n = \sum_{r=1}^{n-1} u_r$, where $u_1 = 1$ and $u_2 = 1$. Prove by induction that $u_n = 2^{n-2}$ for all $n \geq 2$.

Question B: A sequence $u_1, u_2, \ldots, u_n, \ldots$ is defined by the formula $u_n = \sum_{r=1}^{n-1} u_r$, where $u_1 = 1$ and $u_2 = 1$. Find u_3, u_4, u_5, u_6. Make a conjecture for the formula for u_n in terms of n for $n \geq 2$. Justify your answer by mathematical induction.

(Modified from GCE 'A' Level J87/0/1(b))

Figure 3. Two versions of a mathematical induction question

Question A, a standard examination-type question, involves directly getting students to apply mathematical induction to prove the given formula; Question B engages them in looking for patterns, making conjectures and proving their conjecture by mathematical induction. Compared with Question A, this latter task provides a holistic list of student experience in learning mathematical reasoning and will more likely lead them to appreciate mathematical proofs and reasoning.

Teachers should attempt to incorporate more of such mathematical investigations as a skilful means to induce them to formal mathematical proof. More of such questions that link formal mathematical proofs to mathematical investigation (as illustrated in Question B in Figure 3) should be included as essential learning experience for all students learning mathematical induction.

One useful way of engaging students in the reasoning process is to have them examine and explain errors (Carroll, 1999). This strategy, when applied to mathematical tasks, engages the students to think deeply into the mathematical concepts acquired, thereby helping them to regulate their own thinking. Engaging students to critique erroneous and formulate correct mathematical reasoning would also avert the case of students simply accepting formal mathematical proofs as "meaningless rituals" (Toh, 2006). A sample activity is attached in Appendix A, where students are required to identify the erroneous arguments in each question statement and/or proof. Readers should be reminded that this activity in identifying the errors in mathematical proofs is a higher-order thinking task; it should be introduced only after students have achieved a sound conceptual understanding.

3 Connections

According to the mathematics syllabus document (MOE, 2006), "Connections refer to the ability to see and make linkages among mathematical ideas, between mathematics and other subjects, and between mathematics and everyday life. This helps students make sense of what they learn in mathematics" (p. 4).

In this section, sample activities for each of the connections mentioned in the syllabus document: making linkages across (1) mathematical ideas; (2) mathematics and other disciplines; and (3) mathematics and everyday life will be discussed.

3.1 *Making connections across mathematical ideas*

According to NCTM (2000), students from Grades 9 through 12 (age 15 through 18) should "develop an increased capacity to link mathematical ideas and a deeper understanding of how more than one approach to the same problem can lead to equivalent results" (p. 354). The same was also echoed in the Singapore mathematics curriculum described in the preceding paragraph. Connections across different mathematical ideas can be made at the introduction of a topic, as a summary of a main topic or as a typical exercise of a topic. Here, three sample activities to bring out connection among mathematical ideas in teaching A-level mathematics are discussed.

Introducing a new topic

Students perceive some A-level mathematics topics (e.g. Complex Numbers) to be "meaningless" (Toh, 2006), although many of these procedures are manageable. Although the arithmetic processes involved in complex numbers are analogous to those involving surds, it is inconceivable to students why (and how) the number $\sqrt{-1}$ exists.

To fully appreciate complex numbers, one must have a sound understanding of the historical development of complex numbers, which is related to the study of the roots of a cubic equation (see for example, Burton, 2003, pp. 295-302).

Historically, the concepts of complex numbers, quadratic equation, cubic equation, methods of solving a quadratic equation, factor theorem are closely linked. In order to enable students to appreciate complex numbers and demonstrate the connectedness of the various mathematical ideas, a sample pre-introduction activity in teaching complex numbers shown in Figure 4 can be designed for teaching complex numbers.

a. Show that $(a + b)^3 = 3ab\,(a + b) + a^3 + b^3$.

b. You are given the equation $x^3 = 3px + q$. Explain why if two numbers a and b satisfy the equation $ab = p$ and $a^3 + b^3 = 2q$, then $x = a + b$ is a root of the given equation.

c. Show that the roots of the equation in b) can be expressed as

$$x = \sqrt[3]{q + \sqrt{q^2 - p^3}} + \sqrt[3]{q - \sqrt{q^2 - p^3}}\ .$$

(Hint: Note that a quadratic equation can always be expressed in the form of x^2 – (sum of roots)x + (product of roots) = 0.)

d. Use the above formula to find the root of $x^3 = 15x + 4$. Do you obtain

$$x = \sqrt[3]{2 + \sqrt{-121}} + \sqrt[3]{2 - \sqrt{-121}}$$

e. Use factor theorem to find all the roots of $x^3 = 15x + 4$. Do you get three real roots?

f. How do you reconcile parts d and e above?

Figure 4. Sample activity introducing complex numbers

The above activity builds on students' understanding of secondary school mathematics and, more importantly, leads them to appreciate that in fact, the square root of negative numbers could have a "real" meaning! In the subsequent teaching of the roots of a complex number, teacher could build on this activity to identify all the three real roots to the complex number $x = \sqrt[3]{2 + \sqrt{-121}} + \sqrt[3]{2 - \sqrt{-121}}$. The activity in Figure 4 would serve as a useful activity as an introduction to Complex Numbers for interested higher achieving mathematics students; a more qualitative discussion of the underlying history behind the discovery of complex numbers could be used as an introduction to this topic for the less inclined mathematics students.

Readers should note that using elements from the history of mathematics in teaching mathematics is a useful "tool" (Jankvist, 2009) to achieve many pedagogical objectives: it (1) provides a connection across different mathematical concepts (as illustrated above), (2) leads to a heightened awareness and (3) appreciation of mathematics and aids in looking at problems from a different perspective (Bellomo & Wertheimer, 2010). Although history of mathematics is not a "goal" (Jankvist, 2009) in the Singapore curriculum, it is a powerful "tool" (Jankvist, 2009) that mathematics teachers should consider using appropriately in teaching.

Summarizing a big topic

Thematic approach to summarize a huge A-level topic is an approach to help students link several key concepts across the sub-topics together (Toh, 2007). The topic "Sequences and Series" in A-level mathematics consists of several sub-topics: (1) sigma notation; (2) arithmetic and geometric progression; (3) method of difference; and (4) mathematical induction.

One could use a simple mathematics problem that could potentially be linked to many sub-topics, engage the students to brainstorm as many methods as possible to find the solution to the given problem, discuss the challenges, advantages and the implications of each of the methods. As an illustration, consider the following problem:

> Problem: Find the sum of the first n positive integers. Use as many methods as possible. Justify your answer.

Some possible outcomes to the problem are illustrated in Figure 5. The fuller and more in-depth discussion of the various solutions and the implications of each solution are discussed in Toh (2007).

Solving a mathematical problem using different methods spanning from different mathematical topics could help students to see the same problem from different perspectives, thereby appreciating the different mathematics taught at different time. This connection of different mathematical ideas is an important aim of the new mathematics syllabus.

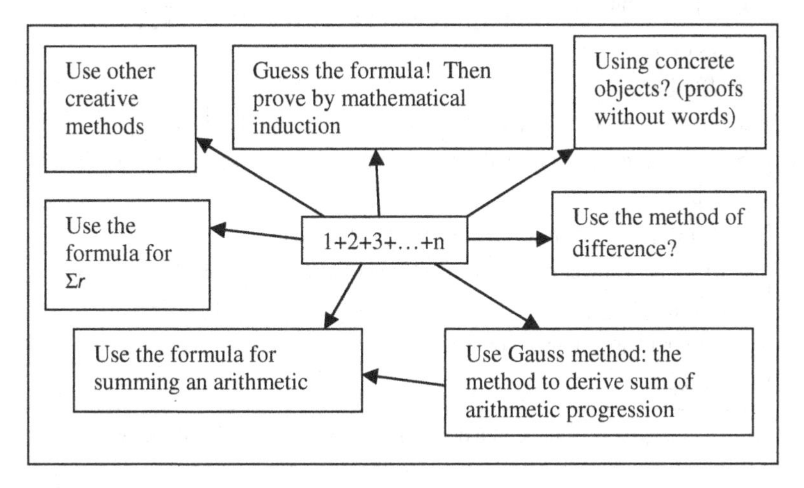

Figure 5. Sample activity connecting different concepts to summarize a large topic

Teachers could use the above type of "summarizing" task as a lecture exercise after they have completed one major A-level lecture topic. Such an activity best serves as a consolidation exercise for all mathematics students.

Tutorial question at the end of a chapter

The use of a mathematically rich question in a tutorial worksheet that connects different concepts is another approach to provide connections across different mathematical ideas. This question could be introduced as a tutorial worksheet problem which requires collaborative effort among several students, but should not be viewed as an "optional" question.

On the other hand, readers are reminded that a question of this nature is generally difficult for average students. As such, the importance of collaborative effort needs to be emphasized. Further, different levels of scaffolding might be required for students of different ability and inclination. An example of a question and the suggested scaffoldings is shown in Figure 6. Teachers could adjust the scaffoldings required depending on their students' capability.

Find the formula for $\frac{1}{1}+\frac{1}{1+2}+\frac{1}{1+2+3}+\frac{1}{1+2+3+4}+........+\frac{1}{1+2+3+...+n}$ in terms of n.

Scaffolding:

a. List the topics that you think this question may be classified.

b. Write down all the formulae that you think may be needed to solve this question.

c. Show how you simplify the question in this step.

d. Examine whether this corresponds to any particular topic that you have learnt.

e. After having obtained your answer, how do you check that your answer is correct?

Figure 6. Sample of a question with the suggested scaffoldings

3.2 *Making connections between mathematics and other disciplines*

Motivation of students is a major problem in teaching mathematics (Kloosterman, 1996; Lambic, 2011). Students do not usually understand the reasons why they want to learn mathematics and how it could be applied (Musto, 2008). It is crucial that teachers use suitable opportunity to connect mathematics with other disciplines that students are familiar with. In this section, two sample activities are discussed in relation to Science.

Teaching of science at A-levels aims to develop students' qualitative understanding of the various scientific concepts. A rigorous understanding of these concepts, usually downplayed in the A-level science curriculum, involves a detailed mathematical treatment. In teaching A-level mathematics, teachers could play an important role to bridge the gap between a qualitative understanding and a rigorous understanding of the science concepts.

Some A-level mathematics topics are predisposed to applications in science subjects; an example being "Differential Equations". Teachers could make use of this opportunity to show mathematically that the half-life of a radioactive decay (a concept that students learn in Physics) is independent of the initial amount of the radioactive substance. Figure 7 provides a sample activity and scaffolding that leads students to achieve a quantitative understanding that the time for the radioactive substance to decay to half of its original amount does not depend on the initial amount of the radioactive substance.

In a radioactive decay, the rate of the decay can be modeled by the equation

$\frac{dx}{dt} = -kx$, where x is the amount of the radioactive substance remaining at time t, and k is a positive constant.

(a) If the initial amount of the substance is x_0, find an expression for x in terms of x_0, k and t.

(b) Find the time for the substance to decay to $\frac{x_0}{2}$, half the original amount.

(c) Does your expression in (b) dependent on x_0?
The half-life of the radioactive substance is the time for a radioactive substance to decay to half its original amount. What does (c) tell you about the half -life of the radioactive substance?

Figure 7. Sample of a question on differential equations that connects mathematics with physics

Consider another A-level topic "vectors". This topic involves many concepts (lines, planes, distance between lines, a line and a plane, etc.). Anecdotal evidence has shown that teachers have taken effort to prepare summary of the various concepts at the end of the topic to facilitate students' learning. The author suggests that teachers could introduce a problem overarching all the various concepts which also introduces another subject discipline as a context for this mathematical problem. A sample question of this type is shown in Figure 8.

The diagram below shows a plane mirror with equation $\mathbf{r}\cdot\begin{pmatrix}1\\2\\3\end{pmatrix}=2$. A ray of light is shone on the plane along the path $\mathbf{r}=\begin{pmatrix}0\\1\\3\end{pmatrix}+\lambda\begin{pmatrix}1\\-1\\2\end{pmatrix}$, where λ is real.

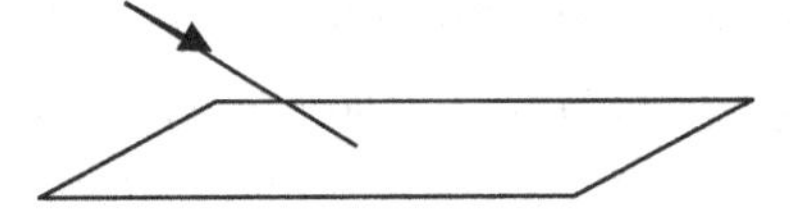

The rules of reflection of light states that: (1) the incident ray, the reflected ray and the normal at the point of incidence all lie on the same plane; (2) the angle of incidence equals the angle of reflection.

a. Explain what you understand by (1) and (2) above. Discuss with your friend beside you (recall what you have learnt in Physics).

b. Find the equation of the path of the reflected ray. (Discuss all the concepts in vectors that are needed to solve this problem)

Figure 8. Sample of a question on vectors that connects mathematics with physics

In Figure 8, students are guided to develop a deeper understanding of the laws of reflection of light and, more importantly, appreciate the application of mathematics in other disciplines. This type of questions could be incorporated as an "optional" question for students who are keen to find out the connection of mathematics to other disciplines.

Mathematics has many applications to other subjects, even the art subjects and aesthetics (for example, Cooper & Barger, 2009). Teachers who are familiar with other disciplines are strongly encouraged to understand how mathematics can be applied to these disciplines and to design appropriate activities that demonstrate the applications of mathematics to other disciplines that are appropriate for their students, to fit the different aptitudes of their students.

3.3 *Mathematics and real-life*

Singapore secondary school mathematics textbooks are replete with many examples to demonstrate the usefulness of mathematics in daily lives, which need not be belaboured at the A-levels. This emphasis on the application of mathematics in real life should be continued at the A-levels.

It is important to get students to appreciate (1) the situation of a mathematical problem in the real-life context; (2) the mathematical solution of a problem might not match exactly the real world solution of the corresponding problem. Here, a sample of such a task is selected from "Differential Equations", adapted from Liu (2003, p. 41).

Part A Find the solution of the differential equation $\frac{dh}{dt} = -k\sqrt{h}$, given that $h = H$ when $t = 0$. Leave your answer for h in terms of k, t and H.

Part B The rate of flow of a viscous liquid out of the cylindrical container (with a leakage in its circular base) may be modeled by the differential equation $\frac{dh}{dt} = -k\sqrt{h}$. The height of the container is H.

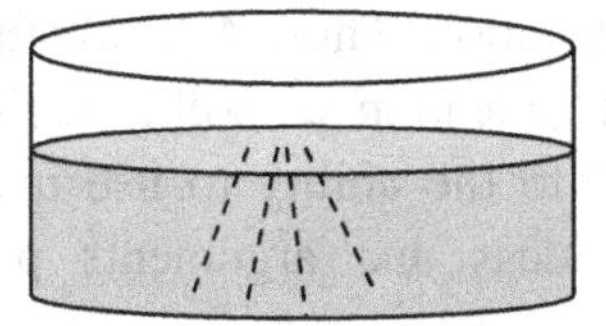

a. Initially, the container is full. Find an expression of h in terms of k, h and H.
b. Sketch a graph of h versus t.
c. Compare your solution to Part A and Part B. What can you say?

Figure 9. Sample of a mathematical problem and a real-world problem for students to compare the two solutions

Part A of Figure 9 is a pure mathematical problem, for which the solution is

$$h=\left(\sqrt{H}-\frac{kt}{2}\right)^2.$$

Part B, which makes use of the differential equation in Part A, is a modeling of the water flow of a leaking container. Interpretation is needed to obtain the full solution of this situation: once all the water has left the container, the height of the water in the container should remain at zero and not increase again! Hence, the solution of Part B is no longer the solution of Part A! It should be

$$h=\begin{cases}\left(\sqrt{H}-\dfrac{kt}{2}\right)^2, & 0\le t\le\dfrac{2\sqrt{H}}{k}\\ 0, & t>\dfrac{2\sqrt{H}}{k}\end{cases}$$

instead. It is important that teachers attempt to connect the mathematical problems to the corresponding real-life situation in their teaching.

4 Communication

Communicating mathematical ideas is gaining increasing emphasis in tertiary mathematics education. Thus, A-level students must be prepared for mathematical communication as well. According to MOE (2006), "Communication refers to the ability to use mathematical language to express mathematical ideas and arguments precisely, concisely and logically" (p. 4).

Opportunities should be given to students in the tutorial lessons to present their solution and communicate their ideas orally. Anecdotal evidence shows that teachers would usually avoid too much student presentation or discussion in tutorial lessons — too much time spent on student presentation or communication would deprive the time necessary for teachers to complete the entire tutorial worksheets. It is important that teachers strike a balance between the teaching time and allocating time for students to present and communicate their solutions.

In tutorial lessons, teachers' use of "What if" and "How" type of questions not only force students to think and reason critically, but allow them the opportunity to present their ideas in an intelligible form and communicate their own ideas and alternative approaches to the same problem. Such effort should be incorporated in the tutorial lessons to enable students to develop their communication ability, as stipulated in the syllabus document.

As an illustration, consider the typical tutorial question:
Solve the inequality $(x + 1)^2 (x - 2)(x - 3) < 0$.

Teachers could build on this question to engage their students to think over how the resulting solution changes accordingly with some small changes in the inequality.

Q: What happens if the strict inequality '<' is replaced by $\leq$?
Q: How will the solution change if $(x + 1)^2$ is replaced by $(x + 1)^3$?
Q: How will the solution change if the inequality is replaced by $\frac{(x-1)^2}{(x-2)(x-3)} < 0$?

Students should be provided with opportunities to communicate their understanding of the related concepts, and how they expect the result would change following a slight change in a given condition of a problem, and how their expectation could be verified mathematically. To promote communication in mathematics classroom, one should not merely focus on discussing the solution of tutorial questions during the lessons.

5 Conclusion

Realizing the ideals of mathematical reasoning, connections and communication is one important step towards developing higher-level mathematical thinking required for problem solving. Generally, classrooms must become environments in which students are able to engage actively in rich, worthwhile mathematical activities (Henningsen

& Stein, 1997). In this chapter, concrete activities together with their rationale on how to infuse mathematical reasoning and connection are discussed. Readers are reminded that feasible suitable tasks to realize these ideals may vary from student to student, depending on their aptitude and inclination; there is hardly any single task that fits all students.

Teachers are strongly encouraged to read up more on how such tasks can be developed, adopted or adapted as classroom activities that bring out the spirit of reasoning, connections and communication in teaching A-level mathematics most suitable for their students. In fact, many standard questions can be easily transformed into mathematical tasks that can engage students in reasoning, communication and connections (Krulik & Rudnick, 1999). This is clearly illustrated by the various sample activities presented in this chapter. Teachers may want to use this chapter as a starter.

References

Bellomo, C., & Wertheimer, C. (2010). A discussion and experiment on incorporating history into the mathematics classroom. *Journal of College Teaching & Learning, 7*(4), 19-24.

Burton, D.M. (2003). *The history of mathematics: An introduction.* New York: McGraw-Hill Publications.

Carroll, W. M. (1999). Using short questions to develop and assess reasoning. In L. Stiff (Ed). *Developing mathematical reasoning in grades K-12* (pp. 247-255). VA, Reston: National Council of Teachers of Mathematics.

Cooper, B.D., & Barger, R. (2009). Listening geometry. *Mathematics Teacher, 103*(2), 108-115.

Healy, L., & Hoyles, C. (2000). A study of proof conceptions in algebra. *Journal for Research in Mathematics Education, 31*(4), 396-428.

Henningsen, M., & Stein, M.K. (1997). Mathematical tasks and student cognition: Classroom-based factors that support and inhibit high-level mathematical thinking and reasoning. *Journal for Research in Mathematics Education, 28*(5), 524-549.

Jankvist, U.T. (2009). A categorization of the "whys" and "hows" of using history in mathematics education. *Educational Studies in Mathematics, 71*(3), 235-261.

Kloosterman, P. (1996). Students' beliefs about mathematics: A three year study. *Elementary School Journal, 97*(1), 39-56.

Krulik, S., & Rudnick, J.A. (1999). Innovative tasks to improve critical and creative-thinking skills. In L. Stiff (Ed.), *Developing mathematical reasoning in grades K-12*, pp. 138-145. VA: Reston, National Council of Teachers of Mathematics.

Lambic, D. (2011). Presenting practical application of mathematics by the use of programming software with easily available visual components. *Teaching Mathematics and Its Applications: An International Journal of the IMA, 30*(1), 10-18.

Liu, H.J. (2003). *A first course in the qualitative theory of differential equations.* New Jersey: Prentice Hall Publication.

Ministry of Education. (2006). *Mathematics syllabus — Secondary.* Singapore: Author.

Musto, G. (2008). Showing you're working: A Project using former pupils' experiences to engage current mathematics students. *Teaching Mathematics and Its Applications: An International Journal of the IMA, 27*(4), 210-217.

National Council of Teachers of Mathematics (2000). *Principles and standards for school mathematics.* VA, Reston: NCTM.

Recio, A. M., & Godino, J. (2001). Institutional and personal meanings of mathematical proof. *Educational Studies in Mathematics, 48*(1), 83-99.

Selden A., Selden, J., Hauk, S., & Mason, A. (1999). *Do calculus students eventually learn to solve non-routine problems?* (Tech. Rep. No. 1999: 5). USA: Tennessee Technological University.

Skemp, R. (1976). Relational understanding and instrumental understanding. *Mathematics Teaching, 77*, 20-26.

Toh, T.L. (2006). Mathematical reasoning from O-level to A-level. *Mathematical Medley, 33*(2), 34-40.

Toh, T.L. (2007). Contextual approach in teaching mathematics: an example using the sum of series of positive integers. *International Journal of Mathematical Education in Science and Technology, 38*(2), 273-282.

Toh, T.L. (2009). On in-service mathematics teachers' content knowledge of calculus and related concepts. *The Mathematics Educator, 12*(1), 69-87.

Toh, T.L., Quek, K.S., Leong, Y.H., Dindyal, J., & Tay, E.G. (2011). *Making mathematics practical: An approach to problem solving.* Singapore: World Scientific.

Appendix A

Comment on each of the following statements and proofs	Comment
1. For any complex numbers $z_1, z_2, \ldots\ldots, z_n$, $\|z_1 + z_2 + \ldots\ldots + z_n\| \le \|z_1\| + \|z_2\| + \ldots\ldots + \|z_n\|$. Proof The result is obviously true for $n = 1$ since $\|z_1\| \le \|z_1\|$ for any complex number z_1. Assume that the result is true for $n = k$, where k is some positive integer ≥ 1, i.e $\|z_1 + z_2 + \ldots\ldots + z_k\| \le \|z_1\| + \|z_2\| + \ldots\ldots + \|z_k\|$. We want to show that the result is true for $n = k + 1$, i.e., $\|z_1 + z_2 + \ldots\ldots + z_k + z_{k+1}\| \le \|z_1\| + \|z_2\| + \ldots\ldots + \|z_k\| + \|z_{k+1}\|$. LHS = $\|z_1 + z_2 + \ldots\ldots + z_k + z_{k+1}\|$ = $\|(z_1 + z_2 + \ldots\ldots + z_k) + z_{k+1}\|$ (treat $z_1 + z_2 + z_3 + \ldots\ldots + z_k$ as one number) $\le \|z_1 + z_2 + \ldots\ldots + z_k\| + \|z_{k+1}\|$ $\le \|z_1\| + \|z_2\| + \ldots\ldots + \|z_k\| + \|z_{k+1}\|$ (since the inequality is true for $n = k$) $\le$ RHS, thus we have completed the proof. Since the inequality is true for $n = 1$, and it is true for $n = k + 1$ whenever true for $n = k$; by induction, it is true for all positive integers n. 2. All the numbers in any set consisting n distinct numbers $\{a_1, a_2, \ldots\ldots, a_n\}$ are equal. Proof Consider a set consisting of one element, i.e. $\{a_1\}$. Clearly, all the elements in the set are equal (since it has only one element). Hence the statement is true for $n = 1$. Assume that the statement is true for $n = k$, i.e. all the numbers in any set consisting k distinct numbers $\{a_1, a_2, \ldots\ldots, a_k\}$ are equal, i.e., $a_1 = a_2 = a_3 = \ldots\ldots = a_k$. Consider a set consisting of $k + 1$ elements, i.e.	

$\{a_1, a_2, ..., a_{k+1}\}$. We can split this into two subsets:
$\{a_1, a_2, ..., a_k\}$ and $\{a_2, a_3, ..., a_{k+1}\}$, both sets containing k elements each. By our assumption for $n = k$, we must have $a_1 = a_2 = a_3 = ... = a_k$ and $a_2 = a_3 = a_4 = ... = a_{k+1}$.
Combining these two results, we must have
$a_1 = a_2 = a_3 = ... = a_k = a_{k+1}$.
Thus, the statement is true for $n = k + 1$ if it is true for $n = k$.
By mathematical induction, this statement is true for all positive integers n.

3. The number e is rational.
Proof We know from power series that

$$e = 1 + 1 + \frac{1^2}{2!} + \frac{1^3}{3!} + ... = \sum_{r=0}^{\infty} \frac{1}{r!}.$$

Let P_n be the statement that " $\sum_{r=0}^{n} \frac{1}{r!}$ is rational".

When $n = 1$, $\sum_{r=0}^{1} \frac{1}{r!} = 1$, which is obviously rational.

Assume that P_k is true for some positive integer k, i.e., $\sum_{r=0}^{k} \frac{1}{r!}$ is rational. Need to show that P_{k+1} is true, i.e. $\sum_{r=0}^{k+1} \frac{1}{r!}$ is rational.

$$\text{LHS} = \sum_{r=0}^{k+1} \frac{1}{r!}$$

$$= \sum_{r=0}^{k} \frac{1}{r!} + \frac{1}{(k+1)!}$$

By P_k, $\sum_{r=0}^{k} \frac{1}{r!}$ is rational, and also that $\frac{1}{(k+1)!}$ is also rational. Since the sum of two rational numbers is rational, we have proven P_{k+1}. By induction, P_n is true for all positive integers $n = 1, 2, 3, ...$ Since it is true for all positive integers, $e = \sum_{r=0}^{\infty} \frac{1}{r!}$ is rational (Proved).

Chapter 8

Visual and Spatial Reasoning: The *Changing* Form of Mathematics Representation and Communication

Tom LOWRIE

Until recently, most mathematics tasks presented to primary-school students were exclusively word-based problems. Current practices, from both curriculum and assessment perspectives, have moved toward more visual and graphic forms of representation. This is unsurprising given the increased use of graphics in society and the ongoing challenge of representing burgeoning amounts of information in visual and graphic forms. From a young age students are exposed to visual forms of communication with more intensity and engagement, whether playing computer games or navigating web pages. This chapter highlights the important role visual and spatial reasoning plays in how mathematics is communicated. In particular, it considers the *changing* nature of mathematics representation in classroom practices, and an evolution in student engagement—where students are increasingly required to decode visual and spatial information. The chapter also considers the need for young students to employ encoding strategies which effectively encourage visual and spatial reasoning.

1 Introduction

It could be argued that mathematics curricula has changed little in recent years and yet the way in which mathematical ideas are represented and communicated have shifted quite dramatically. Until recently, most mathematics tasks that primary-school students were required to solve were heavily word- and number-based whereas the current practice, from both curriculum and assessment perspectives, is to have more graphics embedded into task representation (Lowrie & Diezmann, 2009). This is unsurprising given the increased use of graphics in society and the ongoing challenge of representing burgeoning amounts of information in visual and graphic forms. The amount of information at an individual's disposal and the extent to which this information can be manipulated and directed toward specific purposes has also increased (e.g., the detailed information available for weather forecasts). From a young age children are exposed to visual forms of communication with more regularity and saturation, whether playing digital games, navigating web pages, or interpreting the rich design features of more traditional pictorial representations. Consequently, different forms of sense making are required in order to interpret visual and spatial displays. Nevertheless, the act of engaging with rich visual and spatial displays does not necessarily mean that students can seamlessly understand the information presented to them in these forms. As Metros (2008) argued, being able to view pictures does not ensure a student can create images or that the capacity to read a map ensures an understanding of map data or building a chart better equips a student to depict data in representationally-appropriate ways. It may be the case that specific skills are required to interpret particular graphic displays.

Education systems are paying more attention to the role of representation in school mathematics (e.g., Ministry of Education, 2006). Mathematical representations have always been viewed as an integral component of the ideas and concepts used to understand and engage with mathematics (National Council of Teachers of Mathematics, 2000). However, the structure and purpose of these representations continue to evolve. In this chapter I contend that changes in the way mathematics is "represented" have changed most noticeably through assessment practices. Yet these changes have remained, for the most part, unnoticed.

2 Representations in Mathematics

Representations tend to fall under two systems, namely *internal* and *external* representations (Goldin & Shteingold, 2001). Internal representations are commonly classified as pictures "in the mind's eye" (Kosslyn, 1983) and include various forms of concrete and dynamic imagery (Presmeg, 1986) associated with personalised, and often idiosyncratic, ideas, constructs and images. External representations include conventional symbolic systems of mathematics (such as algebraic notation or number lines) or graphical representations (such as graphs and maps). These two systems cannot and do not exist as separate entities and are seen as "a two-sided process, an interaction of internalization of external representations and externalization of mental images" (Pape & Tchoshanov, 2001, p. 119).

There is, however, some scope (and benefit) of thinking of these two forms of representation in different ways. Internal representations often involve the process of encoding information. Encoding generally occurs when students construct their own representations in order to solve a task. Encoding techniques include drawing diagrams, visualisation and spatial reasoning. These techniques provide students with the opportunity to understand all the elements of any given problem in a way that is meaningful to them. For example, drawing a circle and dividing it into segments in order to better understand a fraction problem. By contrast, decoding techniques are used to make sense of information within a given task, when the information has been represented visually for others to solve (e.g., interpreting a map to determine the coordinate position of a specific street crossing). Ten years ago, a high proportion of mathematics tasks were word-problem based and teachers explicitly taught heuristics which included "draw a diagram", or "imagine the problem scene". These approaches required encoding of information. Currently, a high proportion of tasks have a diagram embedded in the representation. As a consequence, it is hard for students to think beyond the diagram to construct representational meaning and thus, approaches to problem solving now are more likely to require decoding skills.

It is timely to consider the changed nature of mathematics representation in classroom practices, and particularly the evolution of assessment practices—where students are increasingly required to decode information but at the same time less likely to experience situations where they are challenged to encode mathematics ideas and representations.

2.1 *The evolution of assessment tasks*

For many years there has been empirical evidence to support the representation of information in dual forms—that is using graphical or visual displays of information along with text (Kirby, 1993; Paivio, 1971, 1986; Schnotz, Picard & Hron, 1993). Much of this research has focused on the learning of information and how best to represent abstract ideas and concepts to better aid deep understanding. This change in representation has naturally been incorporated within most teaching cycles—from instruction through to assessment. As such, assessment practices worldwide have moved toward more dynamic visual displays in task representation (Lowrie & Diezmann, 2009).

Within Australia, this change is most obvious within the (old) state and (new) national assessment agendas. Prior to the National Assessment Plan for Literacy and Numeracy (NAPLAN) (see for example Australian Curriculum, Assessment and Reporting Authority [ACARA], 2008) being introduced in 2008, Australian states were responsible for skills-based testing—in some states, for nearly 20 years prior. A comparison of two state-based tests in Australia (over a 13-year period) revealed distinct differences between the graphic richness and the worded instructions of each task. Figure 1 highlights the changed nature of the visual displays with the first task (the maze) based on a two-dimensional graphic, relatively free of detail and information that could be considered distractive. The second task (Melanie's model) presents information with both two- and three-dimensional aspects, including an elevation perspective and a bird-eye-view perspective. It is not necessarily that one task is more worthwhile or challenging than the other, but that the processing and mathematics understanding to interpret both tasks are quite different.

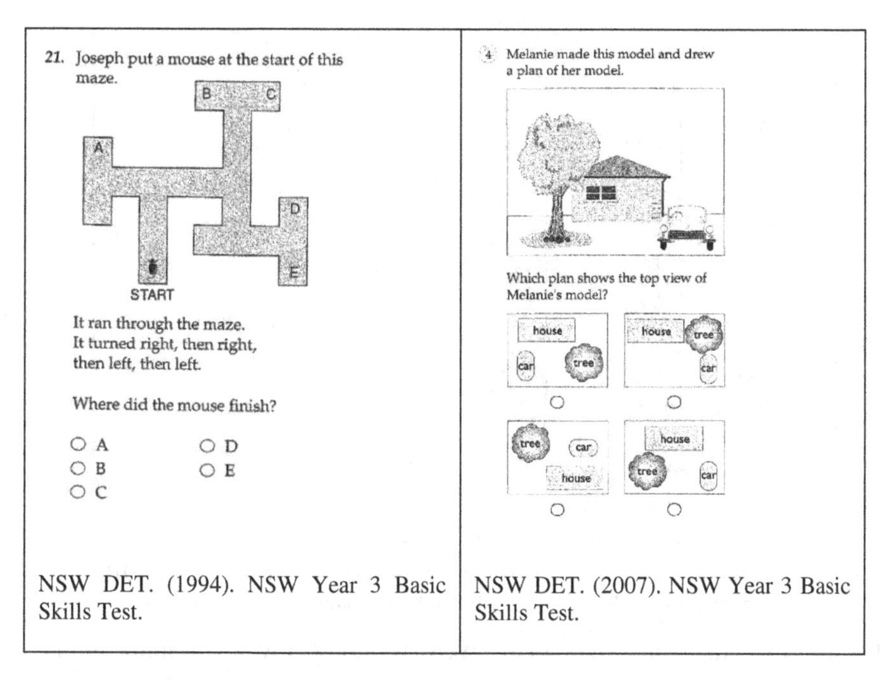

Figure 1. A comparison of two state-based test items 13 years apart.

Not only has the complexity of the graphics changed, but the number of items utilising graphics to (re)present information has increased. Today, graphic tasks which contain information **essential** for task solution are an influential aspect of the assessment "process" in many countries (including Australia and Singapore).

Figure 2 displays the proportion of graphic and non-graphic items found in the NAPLAN (2009-2010) and Singapore Primary School Leaving Examination (Year 6) specimen papers (2006-2010). These instruments are the respective country's national assessments for the primary years. The Singapore specimen papers presented a relatively balanced approach to using graphics in assessment items (59% non graphic items and 41% graphic items), providing students with opportunities to utilise both encoding and decoding skills.

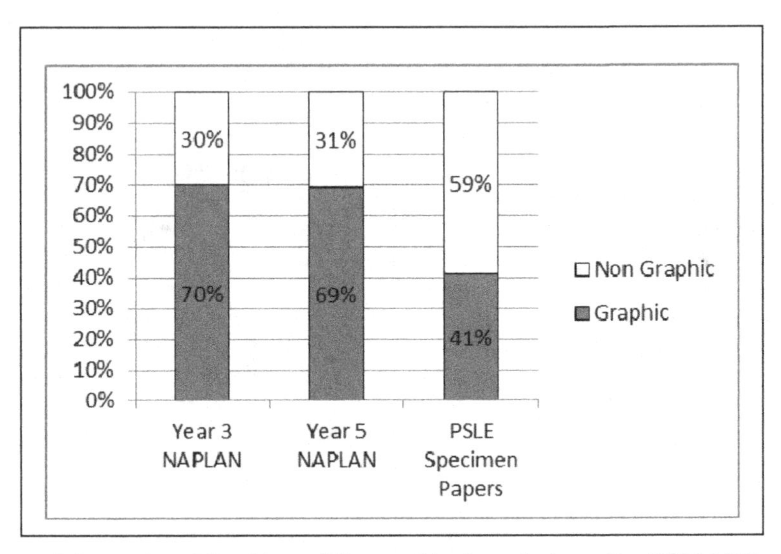

Figure 2. Proportion of Graphics and Non graphics items in Australian (2009-2010) and Singaporean (2006-2010) National Mathematics Assessments.

By contrast, the Australian items were weighted toward graphic items (approximately 70%) across Years 3 and 5. Consequently, students are much more likely to utilise decoding skills (rather than encoding) to solve the assessment tasks.

2.3 *The changing nature of assessment tasks*

It is certainly the case that the structure and nature of NAPLAN-like tasks promote decoding—especially in situations where students are required to generate a multiple-choice solution. Studies (e.g., Lowrie & Diezmann, 2009) have shown that students are reluctant to actually draw on their test booklets when they complete questions in the NAPLAN. Other forms of encoding, including internal representations, are seldom evoked since the answer to the questions generally appear on the page and thus reduces the likelihood of students utilising other forms of imagery. Moreover, the types of questions posed typically require students to decode information from the graphics embedded in the task. By providing a graphical representation to scaffold thinking, a whole new set of skills and practices is brought to the fore. The capacity to interpret various forms of information is now required for students to solve tasks and these skill sets are quite different to those needed when encoding information.

To highlight this case, two probability tasks are presented (Figures 3 and 4), one (re)presented as a traditional word problem and the other designed with text and a graphical display, where information essential to generating a solution is presented in the graphic. Both are valid probability-based tasks. However, they evoke different methods of processing in order to find the solution.

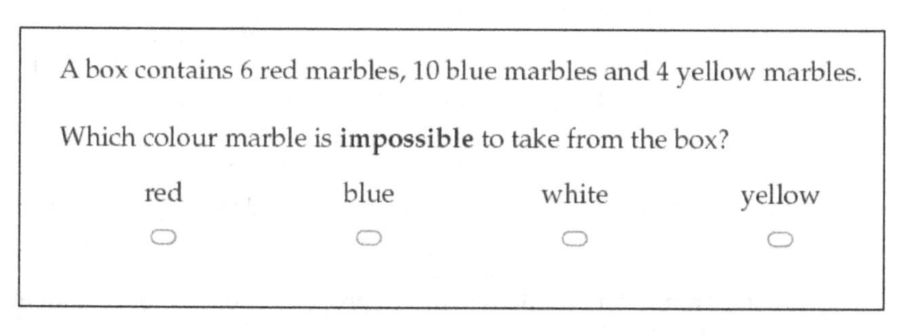

Figure 3. The word-based probability task (ACARA, 2009).

Figure 3 presented students with a relatively basic probability task designed to elicit their understanding of the word *impossible*. Although some young students might use imagery to visualise the box with the marbles in it, or indeed draw the box, for this particular question most students would simply "decipher" that there were no white marbles in the box, hence it is impossible to take one out.

The second task had a similar intent (see Figure 4), that is, to elicit students understanding of the concept *impossible*; only this task was designed using a graphic along with written text. The inclusion of the graphic is meant to provide context as well as information about the type and number of pegs in the bag.

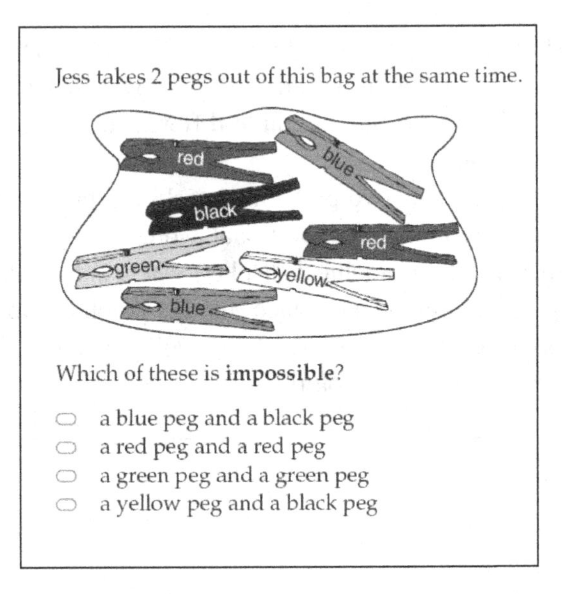

Figure 4. A probability task with a graphic that provides essential information (ACARA. 2010a).

The following transcript highlights the approach taken by one student (Barry).

Barry: I chose a red peg and a red peg…because she couldn't go like that [see Figure 5, where he demonstrates putting his hand into the bag], she would grab the blue (peg on top).

Interviewer: So can I ask you, why couldn't she go down there and get the blue peg (down the bottom)?

Barry: Because the green and the yellow (pegs) are blocking.

Interviewer: Ah, I see. So the black and the red are blocking are they?

Barry: The green and the yellow are blocking the blue, so she would grab the green instead of the blue.

Interviewer: So why did you choose the red and the red?

Barry: Because, see that black and that blue (top peg) they are trying to make it so you can't pull out the red one. So she would pull out, she would go in and pull the blue one out, then she would go into the bag and pull the red one out or the pull out the black.

It would seem that Barry approached this task in a realistic manner—in the sense that he considered the task from an authentic perspective (Lowrie, 2011). The actual representation of the task (Figure 5 shows Barry explaining how he could not get past the pegs at the top of the bag) influenced substantially the approach he undertook. Barry seemed to have a good understanding of the concept of "impossible" however his realistic interpretation of the task precluded him from "seeing" beyond the representation.

Figure 5. Barry explaining how he solved the task.

Such examples demonstrate the challenges that exist when assessment items are framed within realistic contexts. As Cooper and Dunne (2000, p. 43) warned, different types of realistic problems are viewed differently by various social groups and that some students have difficulty "negotiating the boundary between esoteric mathematical knowledge and their everyday knowledge". In this case, Barry was interpreting the task in a realistic manner because the task was set up like as scenario—however, this was not the "true" purpose of the assessment item.

In tasks that establish and contextualise "realistic" situations (like the probability task described in Figure 4), it is important to provide opportunities for students to appreciate the "mathematics purpose" of the task. Thus, for this task, students should be aware that the task is associated with notions of probability. It is easy for students to be distracted by the diagram (and indeed the pegs) rather than conceptual underpinnings of possible and impossible. Asking the following questions can be worthwhile in establishing mathematical purpose but also a sense of task composition.

- What mathematics ideas do I need to use?
- Is the diagram/picture important?
- Could I act this situation out?
- Would it matter if the pegs were arranged in a different order?

These questions allow the students to get some sense of what understandings they need to apply to the task but also promote conversations about why the task was written and presented in a particular manner.

3 Encoding and Decoding Information in Mathematics

With colleagues I have been investigating students' encoding (Lowrie & Logan, 2007) and decoding (Diezmann & Lowrie, 2008; Lowrie & Diezmann, 2007; Logan & Greenlees, 2008) skills as they solve mathematics tasks commonly used as assessment items. The work on encoding has focused on the extent to which students utilise pictures or diagrams to make sense of tasks and the extent to which they evoke imagery to contextualise the problem. The studies which investigate students' decoding skills have considered the extent to which children make sense of information graphics which have different purpose, structure and orientation. One of the current investigations has set out to consider the influence encoding and decoding processes have on primary-school students' mathematical thinking as they complete tasks in the NAPLAN. Year 3 and 5 students (N = 45) who sat the 2010 NAPLAN were interviewed on the 2009 NAPLAN before attempting this year's paper. Students were video-taped as they solved the tasks and explained their solutions to ten items from the respective year NAPLAN tests. The interview protocol encouraged the students to verbalise their thinking and to represent their thinking in ways they felt appropriate (i.e., writing down numbers or drawing a picture). The semi-structured interview allowed students the opportunity to reflect upon an experience that is otherwise only a quantitative measure of performance.

3.1 *Encoding assessment tasks*

The approach taken to solve a task is often influenced by task complexity and novelty, and the various cognitive and affective preferences of the problem solver. Approaches that allow the problem solver to "fold back" (Pirie & Kieren, 1994) to more visual and concrete approaches are most appropriate when difficult or novel tasks are encountered. Task representation can also influence the extent to which students utilise decoding processes to solve tasks—with traditional word problems one type of representation which lends itself well to such processing.

Encoding techniques support students when they do not have the capacity to utilise more analytic processing. For example, in Figure 6, the student drew boxes to represent the cakes and enclosed each group of five circles with a square to represent a box. He then proceeded to keep a tally (in his head) of the number of "cakes" he had represented until he reached 34. He then argued that 7 boxes were required.

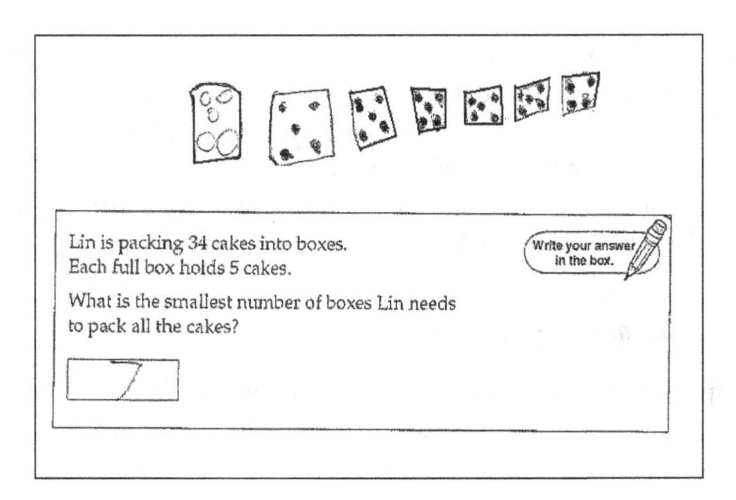

Figure 6. Example of a student using an encoding technique (ACARA. 2009).

This type of procedure represents a common encoding technique utilised by students to solve word problems. As Jack described:

Jack:	I went and drew a box and then I put 5 cakes. And then I drew another box and put 5 cakes and then I drew another one and then another one, another one and another one. But the last one when I drawed (sic) the box I just put 4 cakes in.
Interviewer:	Why did you just put 4 cakes in the last box?
Jack:	Because that was 34.

It is likely that Jack was not able to readily employ an analytic process to solve the task. Rather than using division [$34 \div 5 \cong \square$] or multiplication [$5 \times \square \cong 34$] facts to generate a solution, Jack relied on more primitive visual strategies to represent the task (see Figure 6). Nevertheless, the encoding process that Jack used allowed him to scaffold and utilise information to work out an appropriate solution pathway. It is important that this type of visual processing is encouraged in the mathematics classroom and as educators we need to ensure that such approaches remained valued.

One way of doing this is to encourage students to share with others the approaches they used to solve a task. In order to facilitate learning it is worthwhile to match students who used encoding techniques with students who used analytic methods. Questions students can pose to one another could include:

- Why was a diagram helpful? Which aspects of the diagram helped the most? Confusing?
- Could it have been drawn another way?
- How did it help you make sense of the task?
- Did it help you check (verify) your solution?
- How did you solve the problem without a picture?

Some assessment items provide such scaffolds for students. In Figure 7, for example, the actual graphic within the task is not necessary for task solution and therefore has been displayed, we can presume, to help students contextualise the task.

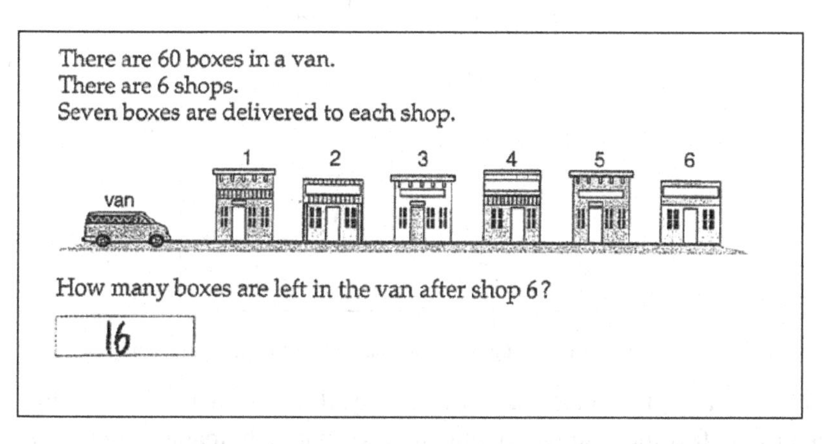

Figure 7. A contextual graphic not essential to task solution (ACARA, 2010a).

I propose that students who have a more sophisticated understanding of such problems would generate an analytic solution with little reference to the graphic in the task. For example, an analytic solution might be 60 – (6 x 7) = □. However, students who find this task very challenging need to "fold back" to more idiosyncratic visual representations in order to support their understandings and make sense of the task. Some of these students would use the graphic provided to help allocate boxes to shops (i.e., adding 7 each time) and then subtracting the total from 60. A more primitive approach is displayed in Figure 8. Gabby represented the task in pictorial form that was certainly laborious. Initially, she represented the 60 boxes as strokes at the top of the page. Gabby then drew 6 circles to represent each shop (in the middle of the page). She then used one-to-one correspondence strategies to mark off each box (from the top of the page) with a stroke under the circle (the shop) until each circle had 7 strokes (or boxes).

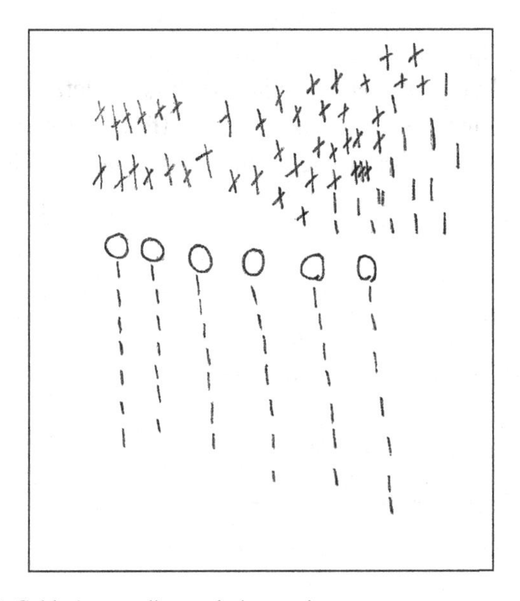

Figure 8. Gabby's encoding technique using one-to-one correspondence.

Unfortunately, this laborious strategy resulted in an incorrect solution due to the fact that she was unable to systematically manage and monitor these procedures accurately. In this instance Gabby would have been better served using the graphic in the task to help represent and keep track of her thoughts more effectively. In this way, a combination of encoding and decoding strategies could co-exist.

From an assessment perspective, students' use of self-drawn diagrams or pictures can provide a reliable guide to their understanding of the task. Thus, while encoding is important strategy for students to utilise in novel or complex situations, it is also a sign of the student's readiness to solve more complex tasks within a certain area. Students who can complete the tasks in an analytic manner typically do not require such scaffolds and consequently demonstrate a different level of capability.

3.2 *Decoding assessment tasks*

As highlighted in Figure 2, current assessment practices require students to undertake a considerable amount of decoding when solving mathematics tasks. Table 1 describes an analysis of Australian students' success when solving assessment tasks that require decoding. The table partitions students' responses across two years of NAPLAN testing in Year 3 and Year 5. The proportion of tasks that contain an information graphic is presented and the number of incorrect solutions by year level and year for the total number of graphic items in each test. The final column reports the proportion of incorrect solutions where an inability to interpret the graphic was the major cause for error.

Table 1

Proportion of incorrect solutions on graphics tasks due to decoding errors

NAPLAN booklet	Proportion of graphic items	Student attempts on graphic items	Incorrect solutions	Proportion of incorrect solutions due to graphic interpretation
Year 3 2009	23/35 (67%)	155	58 (37%)	32/58 (56%)
Year 3 2010	26/35 (74%)	226	79 (35%)	37/79 (47%)
Year 5 2009	26/40 (65%)	115	38 (33%)	16/38 (42%)
Year 5 2010	29/40 (73%)	233	64 (27%)	26/64 (41%)

As previously noted, there were a high proportion of graphic tasks in each of the tests. In fact, both 2010 examinations comprised items containing an information graphic in almost three-quarters of the test. A second level of analysis produced the proportion of incorrect responses that resulted from students being unable to decode the graphic in the respective tasks. These data highlight specific decoding limitations rather than deficiencies in students' "general" mathematics understandings or content knowledge. In other words, instances when an incorrect solution was a direct result of not understanding the graphic rather than an inability to calculate an algorithm; convert equivalent measurements; or understand a particular concept (e.g., notion of "impossible").

By way of example, the assessment task in Figure 9 contains an information graphic which is an essential component of the task. There are a number of identifiable reasons for students producing an incorrect solution on this task. The students could make a calculation error (e.g., 2 x 2 x 4) or make a conceptual error (e.g., not know that the formula for volume is length x width x depth). Such errors would not be classed as misinterpreting the graphic. By contrast, some students could mistakenly misinterpret the graphic and assume that the depth of the carton is only 3 boxes high rather than the intended 4 boxes high. Another way of misinterpreting the graphic would be to incorrectly count all the "imagined" boxes required to fill the carton. Such errors are considered to be decoding errors.

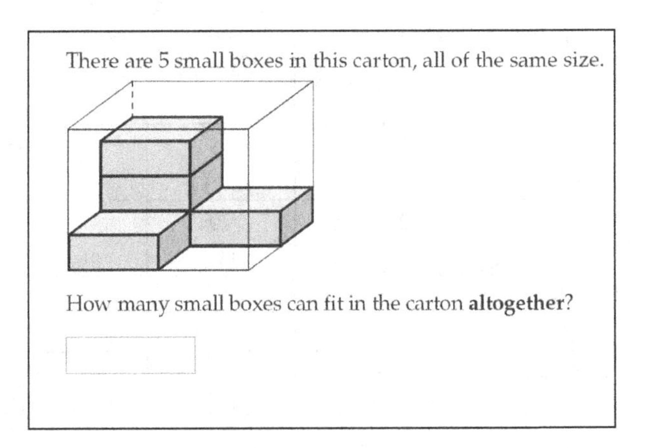

Figure 9. An example of a task which requires decoding using spatial reasoning and mental imagery (ACARA, 2010b).

Thus, the data reported in Table 1 not only shows the increased influence graphics have in assessment tasks, it also reveals the extent to which students are equipped to solve such tasks. It is certainly the case that graphics are an essential ingredient to mathematics skill development. In both Years 3 and 5, more than 40% of incorrect solutions were a result of students not decoding the graphic in appropriate ways. These results provide a compelling case for more explicit teaching of graphics in the mathematics classroom—particularly in the primary school years.

Teachers should encourage students to develop their understandings about graphics in various mathematical situations. Increasingly, we should provide opportunities for students to:

- identify the intent of the graphic and determine the extent to which the information in the graphic is critical for task solution;
- understand the purpose of particular graphics and that some graphics are more appropriately displayed in certain situations than others;
- develop techniques for recognising multiple sources of information in a given task (including text, symbols, keys, legends, axes, and labels); and
- experience visually diverse examples of the same type of graphic to ensure more complex forms of information can be decoded with developmental growth.

Although these techniques should not be considered in isolation, the unique features of graphics needs to be well understood by students to enable them to access and use mathematics that is presented visually.

4 Implications

Several practical implications emerge from the study.

- The movement away from traditional word-based problem solving limits students' opportunities to utilise encoding techniques to make sense of mathematics ideas. If these encoding skills are not encouraged and promoted elsewhere, students' general reasoning skills will be restricted since such techniques are necessary when students encounter novel or complex problems.
- Conversely, the introduction of mathematics tasks rich in graphics requires a different skill base. Explicit attention needs to be given to specific types of graphics since they have different

structure and conventions. Teaching map-based graphics, for example, requires different approaches and techniques than graph-based graphics.

- Given the increasing reliance of graphics in society, it is not surprising that graphic representations hold a prominent place in *current* forms of assessment. And since assessment tends to influence and even drive practice, the way in which mathematics ideas and conventions are represented impact greatly on teaching practices and student learning.
- Students are required to decode external representation with more regularity than the process of evoking internal representations through encoding. Although both require high levels of spatial reasoning, most representations are now "teacher" generated rather than student constructed.
- Students need to acquire different spatial-reasoning skills which allow them to consider all the elements of a task, including specific features of a graphic and the surrounding text, when solving mathematics tasks.
- At least in the current assessment climate, students' performance on national tests need to be considered with caution since overall performance may have a lot to do with students' ability to decode the graphics embedded within tasks.

Acknowledgement

I would like to particularly thank Tracy Logan for her insights and advice in preparation of this manuscript including her analyses of some of the research data.

References

Australian Curriculum, Assessment and Reporting Authority [ACARA]. (2008). *National assessment program literacy and numeracy: Numeracy Years 3 & 5 2008*. Sydney: Author. Retrieved 6 February, 2010 from http://www.naplan.edu.au/tests/naplan_2008_tests_page.html

ACARA. (2009). *National assessment program literacy and numeracy: Numeracy Year 3 2009*. Sydney: Author.

ACARA. (2010a). *National assessment program literacy and numeracy: Numeracy Year 3 2010*. Sydney: Author.

ACARA. (2010b). *National assessment program literacy and numeracy: Numeracy Year 5 2010*. Sydney: Author.

Cooper, B., & Dunne, M. (2000). *Assessing children's mathematical knowledge: Social class, sex and problem solving*. Buckingham, UK: Open University Press.

Diezmann, C. M., & Lowrie, T. (2008). Assessing primary students' knowledge of maps. In O. Figueras, J. L. Cortina, S. Alatorre, T. Rojano & A. Sepúlveda, (Eds.), *Proceedings of the Joint Meeting of the International Group for the Psychology of Mathematics Education 32, and the North American chapter XXX* (Vol. 2, pp. 415-421). Morealia, Michoacán, México: PME.

Goldin, G., & Shteingold, N. (2001). Systems of representations and the development of mathematical concepts. In A. A. Cuoco (Ed.), *The roles of representation in school mathematics* (pp. 1-23). Reston, VA: National Council of Teachers of Mathematics.

Kirby, J. R. (1993). Collaborative and competitive effects of verbal and spatial processes. *Learning and Instruction, 3*, 201-219.

Kosslyn, S. M. (1983*). Ghosts in the mind's machine*. New York: Norton.

Logan, T., & Greenlees, J. (2008). Standardised assessment in mathematics: The tale of two items. In M. Goos, R. Brown & K. Makar (Eds.), *Navigating currents and charting directions* (Proceedings of the 31st annual conference of the Mathematics Education Research Group of Australasia, vol. 2, pp. 655-658). Brisbane, QLD: MERGA.

Lowrie, T. (2011). "If this was real": Tensions between using genuine artefacts and collaborative learning in mathematics tasks. *Research in Mathematics Education, 13*(1), 1-16.

Lowrie, T., & Diezmann, C. M. (2009). National numeracy tests: A graphic tells a thousand words. *Australian Journal of Education, 53*(2), 141-158.

Lowrie, T., & Diezmann, C. M. (2007). Solving graphics problems: Student performance in the junior grades. *The Journal of Educational Research, 100*(6), 369-377.

Lowrie, T., & Logan, T. (2007). Using spatial skills to interpret maps: Problem solving in realistic contexts. *Australian Primary Mathematics Classroom, 12*(4), 14-19.

Metros, S. E. (2008). The educator's role in preparing visually literate learners. *Theory into Practice, 47*(2), 102-109.

Ministry of Education. (2006). *Mathematics syllabus - Primary.* Singapore: Author.

National Council of Teachers of Mathematics. (2000). *Principles and standards for school mathematics.* Reston, VA: Author.

New South Wales Department of Education and Training (NSW DET). (2007). Basic skills test: Aspects of numeracy. Sydney, Australia: Author.

New South Wales Department of Education and Training (NSW DET). (1994). Basic skills test: Aspects of numeracy. Sydney, Australia: Author.

Paivio, A. (1971). *Imagery and verbal processes.* New York: Holt, Rinehart & Winston.

Paivio, A. (1986). *Mental representations: A dual coding approach.* New York: Oxford University Press.

Pape, S. J., & Tchoshanov, M. A. (2001). The role of representation(s) in developing mathematical understanding. *Theory into Practice, 40*(2), 118-127.

Pirie, S. E. B., & Kieren, T. E. (1994). Growth in mathematical understanding: How can we characterise it and how can we represent it? *Educational Studies in Mathematics, 26*(2–3), 165–190.

Presmeg, N. C. (1986). Visualisation in high school mathematics. *For the Learning of Mathematics, 6*(3), 42-46.

Schnotz, W., Picard, E., & Hron, A. (1993). How do successful and unsuccessful learners use text and graphics? *Learning and Instruction, 3*, 181-199.

Chapter 9

Understanding Classroom Talk in Secondary Three Mathematics Classes in Singapore

David HOGAN   Ridzuan Abdul RAHIM   Melvin CHAN

Dennis KWEK   Phillip TOWNDROW

This chapter discusses the interactive opportunities for learning that teachers provide to their students through one key element of classroom talk — teacher questions. We study three broad categories of teacher questions: performative, procedural and conceptual. We also subdivide conceptual questions into clarifying, connecting, explanatory and epistemic questions. Using data from a survey of over 1,000 Secondary 3 students in 30 schools drawn from a representative random stratified sample of secondary schools in Singapore, we report on: 1) the kind of teacher questions that teachers focus on in mathematics classrooms; and 2) the relationships between different kinds of teacher questions. The relative prevalence of performative questions arising from the highly prevalent and mundane triadic talk structure: teacher initiate (I), student respond (R) and teacher evaluate (E) [IRE]suggests that Singapore mathematics classrooms provide limited opportunity for students to engage in rich classroom conversations. However, deeper structural analyses of teacher questions indicate that not all performative questions lead to a *cul de sac*, and that a substantial proportion of them lead on to procedural and explanatory talk. We think this helps explain the success of Singaporean students in international assessments despite relatively low levels of classroom dialogue. It also suggests that we need to be extremely cautious in judging "closed" teacher questions and performative talk, including IRE, as a hum-drum ritualized exchange that has little or no capacity to prompt students to work on understanding.

1 Introduction

In a classic essay, Douglas Barnes (1992; 2008) pleaded for a discursive regime in classrooms focused on "exploratory" talk that "worked on understanding" rather than testing to see whether students knew the right answer or not in performative "presentations." But among contemporary educational researchers, Professor Robin Alexander from Cambridge University is particularly noteworthy for his efforts over many years to theorize the overall importance of talk in the classroom, to develop and explain at both a theoretical and practical level a normative model of high quality classroom talk that he terms "dialogue" and to undertake a series of exemplary qualitative studies of classroom talk:

> Of all the tools for cultural and pedagogical intervention in human development and learning, talk is the most pervasive in its use and powerful in its possibilities. Talk vitally mediates the cognitive and cultural spaces between adult and child, among children themselves, between teacher and learner, between society and the individual, between what the child knows and understands and what he or she has yet to know and understand. Language not only manifests thinking but also structures it, and speech shapes the higher mental processes necessary for so much of the learning that takes place, or ought to take place, in school...
>
> It follows that one of the principal tasks of the teacher is *to create interactive opportunities* and encounters that directly and appropriately engineer such mediation.
>
> Yet, though most educators subscribe to this argument in broad terms, and classrooms are places where a great deal of talking goes on, talk that in an effective way and sustained way engages children and scaffolds their understanding is much less common than it should be. Teachers rather than learners control what is said, who

says it and to whom. Teachers rather than learners do most of the talking. And, as many UK and US researchers have consistently found, one kind of talk predominates: the so-called "recitation script" of closed teacher questions, brief recall answers and minimal feedback that requires children to report someone else's thinking rather than to think for themselves, and to be judged on their accuracy or compliance in doing so..." (Alexander, 2008a, pp. 92-93).[1]

In this chapter we report on recent work in what we call the Core 2 project we are undertaking on the nature of, and opportunities for, classroom talk in Secondary 3 mathematics classes in Singapore. Specifically, we will draw on the results of a survey of a nationally representative sample of almost 1200 students in 32 secondary schools across the city in 2010 to report on the nature of the questions that teachers ask students during lessons and the interrelationships between them. In doing so we want to explore what to many appears as a striking paradox that is yet another version of the so-called East Asian paradox: how do we reconcile the fact that given that international research on classroom talk emphasizes the ability of rich classroom conversations ("dialogue") to promote "deep learning" yet Singapore's students do exceptionally well in international assessments in mathematics (TIMSS, PISA) even though there is clear evidence that students have quite limited opportunity to engage in rich classroom conversations in Singapore? This suggests a number of alternative hypotheses, two of which are of special interest to us here: (1) that high quality talk can facilitate, but is not necessary to, high quality learning, and that this is the case in Singapore; or (2), that advocates of the value of "dialogue" such as Alexander have unnecessarily discounted the educational value of less exalted and more mundane forms of classroom talk, including the much maligned "recitation" or IRE. Unfortunately we are not yet able to report on the full range of exchanges that takes place in the classroom (the particular data set we report in this chapter focuses only on the nature of the questions that teachers ask) and the relationship between

---

[1]See also Alexander (2008c)

the modes and quality of classroom talk, the design of instructional tasks and student learning. Instead, that will come later in the life cycle of the research project. However, we are in a position to report now on the nature of the questions teachers ask and the interrelationships between them.

2 Theoretical Framework

Professor Alexander, like other leading researchers of classroom talk before him — Douglas Barnes, Jay Lemke, Derek Edwards, Neil Mercer, Hugh Mehan, Courtney Cazden, Frederick Erickson, Martin Nystrand and his colleagues — noted that in most classrooms he looked in, whatever the country, the typical lesson was likely to be characterized by the recitation script or IRE (Initiate, Response, Evaluate) sequences described initially by Sinclair and Coulthard in 1975, formally identified by Hugh Mehan in 1979 and re-termed "triadic dialogue" by Lemke a decade later (1989; 1990). Two British researchers, Adam Lefstein and Julia Snell (2010, p.167) recently summarized the underlying structure of IRE/IRF sequences in these terms:

> ...over three decades of research in a wide variety of Anglo-American schools have found relatively consistent patterns in the whole class teaching observed... Teachers dominate classroom interaction, talking most of the time, controlling topics and allocation of turns, judging the acceptability of pupil contributions, and policing inappropriate behaviour. Pupils talk much less than the teacher, for shorter durations and in most cases only in response to teacher prompts. Whole class discourse is typically structured in Initiation-Response-Evaluation (IRE) cycles: teachers *initiate* topics, primarily by asking predictable, closed questions that test pupils' recall of previously transmitted information; pupils *respond* with brief answers; and teachers *evaluate* pupil responses, praising correct answers ("well done!") and/or censuring error ("you haven't been paying attention!"). (Some researchers prefer IRF [Initiation-

Response-Feedback] to IRE, thereby signalling the multiple functions that can be performed in the third move).

Similarly, a team of US researchers led by Martin Nystrand et al. (2003) concluded that IRE sequences dominated English and social studies classrooms in the Mid-West in the US:

> ...a large body of empirical work over the last century has documented the predominance of recitation as the principal mode of whole classroom discourse in American elementary and secondary classrooms...Recently, [we]... found that the vast proportion of questions in a large, diverse sample of eighth- and ninth-grade English classrooms are indeed asked by the teacher in recitation, with whole-class discussion (open exchange of ideas) averaging less than 50 seconds per lesson in the eighth grade and less than 15 seconds in the ninth grade... The dominant profile of whole classroom discourse in these classes involved highly codified test questions, which developed little more than procedural (IRE) reciprocity; moreover, coherence from topic to topic was typically weak or absent. In all classes, the most common purpose of classroom discourse was to recall and display assigned information, to report on what was already known (Nystrand, Wu, Gamoran, Zeiser, & Long, 2003, pp. 138-139).

The dominant research view of IRE is that it is both pervasive and less than benign in its educational consequences. In a recent essay, the distinguished Canadian researcher Gordon Wells, for instance, summarizes the conventional wisdom thus: "the recitation script has been criticized from several points of view: it disadvantages children from cultures in which this form of interaction is uncommon; it provides no bridge from everyday registers to those in which disciplinal knowledge is constructed; and it provides little or no opportunity for students to voice their own ideas or comment on those of others" (Wells & Arauz, 2006, p. 380). In effect, for many commentators, IRE sequences have very limited, if any, capacity, to promote student understanding or cognitive depth. Indeed, the primary function of IRE is evaluative or performative rather than exploratory and constructive. Similarly, Lefstein and Snell (2010, p. 170) report the conventional critique of IRE/IRF sequences and "closed questions" in these terms: "First, the structure positions teachers (and textbooks) as the sole legitimate sources of knowledge; the pupils' role is to recall and recite for evaluation what they have previously read or been told. Second, the structure tends to produce a rather disjointed lesson overall, with teachers moving from topic to topic with little or no clear line of reasoning. A third criticism is that, to the extent that participants do engage in more demanding cognitive activities (e.g. explaining concepts, relating ideas to one another, challenging and/or justifying positions), the bulk of the work is performed by the teacher."

Meanwhile, the conventional wisdom holds that "dialogue" is by far and away the most effective means of promoting deep student learning. As Wells and Arauz put the matter, "learning is likely to be more effective when students are actively engaged in the dialogic co-construction of meaning about topics that are of significance to them...[C]oming to know involves much more active participation by learners in which they construct and progressively improve their understanding through exploratory transactions with the cultural world around them" (2006, p. 379). Similarly, Hodgkinson and Mercer recently observed, apropos the conventional wisdom, that "it is now appreciated that classroom talk is not merely a conduit for the sharing of information, or a means for controlling the exuberance of youth; it is the most important educational tool for guiding the development of understanding and for jointly constructing knowledge" (2008, p. xi).

Few researchers have done more to develop this elevated understanding of the educational value of classroom talk and the cognitive power of dialogue than Robin Alexander. For Alexander (2008b, pp. 184-191), "dialogue" is a form of classroom talk that purposively builds student understanding over the course of the lesson in a process that exhibits evidence of *"purposefulness"*, *"reciprocity"* and *"cumulation."* Technically speaking, "dialogic teaching harnesses the power of talk to stimulate and extend pupil thinking and advance their learning and understanding. It helps the teacher more precisely to diagnose pupil needs, frame their learning tasks and assess their progress. It empowers the student for lifelong learning and active citizenship.

Dialogic teaching is not just any talk. It is as distinct from the question-answer and listen-tell routines of traditional teaching as it is from the casual conversation of informal discussion...." (Alexander, 2010, p. 1). Elsewhere, Wolfe and Alexander (2008) distinguish:

> between *conversation* that tends to be relaxed and may lead nowhere and *dialogue*, characterised by purposeful questioning and chaining of ideas into "coherent lines of thinking and enquiry" — the dialogic principle of *cumulation*. This tilts control of the conversational floor away from the teacher's initiating moves to students' responsive utterances, the R in I(R)F. By listening and responding to what children actually say and do, teachers are in a position to support individuals more effectively in their learning, a principle enshrined in formative assessment...and the extended notion of "learning as assessment" [in which] learning is defined not only as acquisition of knowledge but more potently as participation in knowledge building practices' (p. 8).

For Alexander and others then, dialogue, and the process of cumulation within it, is at the heart of "knowledge building," "knowledge generation" or "constructing knowledge." Alexander (2008b, pp. 184-191) suggests that dialogue involves purposively working towards achieving a common understanding or shared meaning through structured, sequenced, chained, cumulative questioning and discussion that expedites the handover of concepts and principles. What is crucial is that the sequence facilitates development of meaning and understanding and deepens or moves the argument, story or explanation forward. Consequently, "cumulation" occurs when "teachers and students build on their own or others ideas and chain the claims into coherent lines of thinking and enquiry." Unless cumulation occurs, classroom talk remains "discussion" rather than "dialogue."

While our evidence from Singapore indicates that teachers employ IREs in a substantial proportion of their lessons and that a substantial proportion of IREs are essentially performative in character, our evidence suggests that not all performative talk is a cognitive cul-de-sac. Indeed, our sense at this point in time is that many IRE sequences, especially in Mathematics, are *constructive* and not just *performative* in character — that is to say, they are not just about testing for student knowledge but very often also about co-constructing a domain-specific narrative (for example, in Mathematics, explaining or solving a problem, or constructing a representation of a problem) and are epistemically productive and cognitively expansive. If we can establish this conclusion, it would go some way to help explain the intricate relationship between classroom talk and knowledge representation at the classroom level, the multi-dimensional character of understanding talk, and the success of Singaporean mathematics at an international level.

We recognize that this view represents a decidedly revisionist take on the educational value of IRE and that it is hardly the first time such views have been expressed. Indeed, as early as 1975, Sinclair and Coulthard more or less endorsed IRE-type exchanges as a reasonable enough default form of classroom discourse. A decade later, J. L. Heap (1985), thinking grandly, suggested that triadic dialogue was essential for the co-construction of cultural knowledge. In 1992, Neil Mercer concluded that IRE sequences were useful for "monitoring children's knowledge and understanding," "guiding their learning," and "marking knowledge and experience which is considered educationally significant or valuable." A year later, Gordon Wells (1999, p. 3) [2] expressed misgivings about the bad press that IREs were getting among educational researchers, and that they needed to be a little more discriminating and nuanced in their evaluation of IRE sequences. Taking his cue from the work of Nystrand (1997) and Nystrand and Gamoran (1991) in the United States, Wells (1999, pp. 2-4) wrote that it was important to differentiate the different kinds of purposes that IRE exchanges sequences might serve: "...in the hands of different teachers, the same basic discourse format can lead to very different levels of student participation and engagement. It can also be used by the same teacher in different contexts, to achieve very different purposes." Consequently, he goes on, "in itself, triadic dialogue is neither good nor bad; rather, its merits — or demerits — depend upon the purposes it is used on particular occasions, and upon the larger goals for which these purposes

[2]See also Nassaji and Wells (2000)

are informed" (1999, p. 169). Or, as Courtney Cazden (2001) points out, the cognitive impact of teacher questions is affected by how the teacher optimally places particular questions during the classroom exchange, thereby opening up the range of potential responses. Indeed, with respect to all questions, "there is a crucial difference between helping a child somehow get a particular answer and helping that child gain some conceptual understanding from which answers to similar questions can be constructed at a future time" (p. 92).

We agree with this judgment, and endeavour to support a closely related argument with Singaporean data focused on the kinds of questions teachers ask their students. This is a matter of considerable theoretical and practical interest, given that teacher questions are the critical gatekeeper of the character and quality of the classroom exchanges that follow. However, in this chapter we will not attempt to link the pattern of teacher questions, as a proxy for the structure of classroom talk more generally, to the explicit goals or purposes teachers have, as Wells suggests. Instead, we will draw on relatively decontextualized survey data from Secondary 3 students to model statistically the causal relationships between different kinds of questions that teachers ask their students. In later papers we will, however, attempt to fill out the picture by examining the oral exchanges that follow-on from teacher questions (and indeed, from student questions to teachers and other students) and by linking the pattern of classroom talk to the organization of lessons that structure opportunities for student participation in oral exchanges with their teacher or each other (Barnes, 2008; Galton, 2009; Mercer & Littleton, 2007; Hattie 2009; Hogan, Chan, Rahim, Luo, Sheng, Khin, Towndrow & Kwek, 2012), to key epistemic and cognitive characteristics of the instructional tasks that teachers ask students to engage using student survey data and data derived from our coding of 625 video-graphed lessons and transcriptions of a purposeful subsample of this larger sample of 625 lessons, and to a broad array of student outcomes (Hogan, Towndrow, Rahim, Chan, Luo & Sheng, 2011). For now though our focus is much more limited and decontextualized — the relative frequency of three general kinds of teacher questions — performative, procedural and conceptual — and their interrelationships, drawing on the Secondary 3 student survey data.

Discourse analysts have long been interested in identifying what kinds of questions teachers ask in class during a lesson, when, how often, and with what consequences, discursive or otherwise. This reflects three preoccupations. First, as we noted above, is that while IRE exchanges in classrooms are pervasive, teacher initiated questions are often still seen to hold the potential to promote higher order thinking, reflection, self-examination and enquiry, particularly through the use of questions which invite students to speculate, reason, evaluate, conceptualise, or consider a range of possible answers on the grounds that "open" rather than "closed" questions promote co-construction of knowledge and conceptual development (Brown & Wragg, 1993; Galton, Hargreaves, Comber, Wall & Pell, 1999). As Cazden (2001, p. 92) argues, "at their best, teacher questions can both assist and assess student learning." Second, researchers recognize that teacher questions are a fundamental feature of the political and epistemic economy of the classroom, propping up the asymmetrical authority relations in the classroom in which what is learned and valued is controlled by the teacher or, alternatively, creating opportunities for students to participate dialogically as members of a learning community. All too often, however, the teacher exercises linguistic dominance in the classroom and manipulates classroom discourse without regard to student learning (Barnes, Britton & Torbe, 1986). This involves not just managing turn-taking and maintaining order but also determining which student contributions are to be valued and ratified as valid knowledge for the class (Dillon, 1994). The third preoccupation follows from the first two: Unlike most contexts found outside the classroom, the teacher asks questions to which he or she already knows the answer. Students who quickly learn the norms of classroom discourse know that there is usually one answer, or a limited range of answers that are acceptable. For this reason, Postman and Weingartner (1969) call such questions "guess what I'm thinking" questions. A consequence of this is that the process of questioning quickly acts to align students' thinking with the teacher's: it is less a process of educational enquiry but more a process of following the teacher's script (Francis, 2002). When teacher questioning can potentially construct the teacher-student relationship and teaches students that their own knowledge is subordinate to the teacher, challenges immediately arise in how higher order thinking and learning can occur in the classroom discourse (Hardman, 2008).

The empirical evidence on teacher questioning in classrooms internationally, and across the past three decades, supports the finding that questions often narrowly construe the potential space for substantive knowledge work. Barnes et al.'s (1986) analysis of teachers' questions famously drew attention to the disproportionate number of questions that require a predetermined answer — "closed", known-answer, test or display questions. In contrast, "open" questions (variously called genuine, authentic, information-seeking) do not seek one right answer or a limited range of acceptable answers, with the teacher sometimes not having any predetermined answer in mind. Such question types are found to be rare in Barnes et al.'s research. They conclude that teachers use questions to limit thinking to factual recall, rather than using questions to develop learning and understanding. Robin Alexander's (1992) study reaffirms this finding and he argues that while questions were a prominent feature in the teachers' pedagogical repertoire, they rarely exploited the full potential of questioning as a teaching strategy. Likewise, Galton's ORACLE project, conducted in 1976 and replicated in 1996, reported the dominance of closed questions in both periods, and concluded that despite curriculum reforms, the underlying pattern of discourse remains one where teachers talk and students sit and listen (Galton, Hargreaves, Comber, Wall & Pell, 1999). Myhill and Dunkin (2005) paint a similar picture of how teachers use questioning in whole class contexts to maintain control and support their teaching rather than promote student learning. Categorising teacher questions used in literacy and numeracy lessons into factual, procedural, speculative, and process types, they found that while both curriculum areas contain a dominance of factual questions, there are significantly more process questions in numeracy lessons, whereby teachers asked questions which invited students to explain their thinking ("how did you work that out?," "how do you know that?," "can you explain why?"). Their study showed that the two most common function of questions in numeracy lessons are factual elicitation and to practise skills. Myhill and Dunkin go on to advocate that "there is a pressing need to develop pedagogic confidence in framing discourse which permits children to be 'active in creating their own understanding' … and which breaks free of the routinized teacher-child-teacher-child interaction patterns" (2009, p. 426).

We can think of no good reason to challenge the current research understanding of the educational value of open questions and classroom dialogue. Nor do we want to challenge the argument that IREs typically constrain discursive opportunities and the co-construction of knowledge and understanding. What we want to do in this chapter is to focus on the nature and interrelationships between teacher questions using student survey data to ask whether performative questions necessarily close down classroom exchanges or whether they can open up classroom dialogue in ways that promote "knowledge talk" in the classroom. Plainly, not all forms of classroom talk are forms of knowledge talk. Indeed, some forms of non-knowledge talk are unavoidable — organizational talk, and classroom management talk, comes especially to mind. Other forms of classroom talk are weak forms of knowledge talk but still useful for engaging students, including "sharing talk" in which students share experiences or understandings but do not build on each other's understandings. The strongest form of knowledge talk, and the form of knowledge talk that we are most interested in theoretically, is "understanding talk" where we take this to be talk that involves "working on understanding." The conventional research view though is that all too often knowledge talk in classrooms does not focus on understanding. Rather, most of it is factual, expository, asymmetrical, routine and performative, with precious little attention to exploring ideas or "working on understanding." In dialogical forms of understanding talk, on the other hand, students and teachers work together to clarify the meaning and ideas, offer and explain reasons, invite and discuss alternative perspectives and solutions, make connections and establish conceptual relationships, justify beliefs, frame and reframe arguments, and, optimally, engage in rich dialogical exchange characterized by what Robin Alexander terms "reciprocity" and "cumulation." However, we are not convinced that understanding talk, or more pointedly, working on understanding, *only* happens when dialogical talk occurs. Indeed, we suspect that conceptual understanding might well develop in the absence of dialogical exchanges — that is, that dialogical teaching is not *necessary* for understanding, although it might well be especially efficacious in doing so. Critically, while dialogical talk (in the sense intended by Alexander) is a form of understanding talk, understanding talk need not necessarily be dialogical in character. Instead, what is crucial about understanding talk is that it focuses on student meaning making and developing understanding whether or not it involves dialogical exchanges. When such exchanges exhibit, over time, purposefulness, reciprocity, criticality, meaningfulness, and above all, cumulation, we also have evidence of dialogical exchanges, although we believe that it is not possible to establish this until the end of the lesson.

Unfortunately, we cannot test this broader hypothesis directly now. But we can certainly establish whether or not different kinds of teacher questions are covariant or orthogonal to each other. This would allow us to go some way to establishing whether or not "working on understanding" happens under quite diverse discursive regimes, including ones that are conventionally considered inhospitable to the development of understanding. This is not to challenge the claim that dialogical teaching is especially productive of understanding — rather, it is simply to say that while it might well be that dialogical teaching is especially productive for understanding, and even sufficient for it to occur, it is arguably not necessary. But we will be able to report, however, that performative questions are very often followed up by conceptual and procedural questions that facilitate students, to use Douglas Barnes's phrase, "working on understanding" (Barnes, 2008, ch.1). While these might not lead to what Nystrand et al. (2003) term "dialogical spells," let alone "dialogical teaching," our evidence suggests that performative questions, at least in the Singapore context, are quite often followed by questions that look for all the world like they are intended to prompt students to "work on understanding."

3 Mapping Teacher Questions in Singapore

In a widely cited study of classroom interaction in English classrooms conducted in 1976/77, Galton, Croll and Simon (1980, pp. 86-87) reported that 90.4% of all curriculum related questions (as opposed to classroom management questions) were closed questions in that they required a straightforward factual answer or a correct solution to a problem, while just 9.6% were "open" questions in which more than one response was acceptable. In a follow up study in 1996, Galton et al. (1999) reported 86% of all curriculum related questions teachers asked were "closed" and only 14% "open." Alexander (1992) and Smith and colleagues (2004) have produced very similar results with different groups of English primary teachers and at different historical moments.

Table 1 reports the results of a transcript analysis conducted by Doyle and Hong (2009) of teacher questions in Primary 5 and Secondary 3 mathematics in a large representative sample of Singaporean classrooms in 2004 and 2005. As we can see, an overwhelming majority (96.3%) of the questions that teachers asked were "closed" rather than "open" questions — that is to say, they were questions that simply asked students to give a brief answer to a question ("what is the product of 7 multiplied by 9.5") where the answer is already known by the teacher rather than questions that allowed or prompted students to think aloud on their feet, speculate, connect, explain or in some other way offer a more discursive answer to the question. In effect, Doyle and Hong estimated that almost 97% of the questions teachers asked were what we term "performative" rather than productive or generative.

Table 1
Open and closed questioning, primary 5 and secondary 3, 2004

Question Types	Open		Closed		Total questions
	No.	%	No.	%	
Primary 5 Mathematics	111	2.3	4,774	97.7	4,885
Secondary 3 Mathematics	119	6.1	1,183	93.9	1,302
Total	230	3.7	5,957	96.3	6,187

Now, this obviously paints a pretty bleak picture. But we also think it is somewhat misleading as well — not that the numbers are wrong, but the inferences or implications that we might draw from them might not tell the whole story. We suspect, for example, that teacher responses to student responses (the "E" or "F" in the IRE/IRF sequence) might well be just as critical as the initiating question (the "I" in the IRE/IRF sequence) in determining the overall character and quality of the exchange between teachers and students, and whether the exchange can fairly be said to result in students "working on understanding." Indeed, in our coding of classroom interactions in a related project, we are coding for exactly this possibility. But beyond this, we also hypothesized that we can use the student survey data to demonstrate that performative questions are not necessarily dead ends, cognitively speaking, and that they can in fact (although not always, or even very often) promote forms of classroom talk that enables students to "work on understanding" and therefore approximates what we have termed "understanding talk" where we understand this to mean talk that supports conceptual development and complex forms of communication. Obviously, what we might term "conceptual talk" is especially likely to satisfy these conditions, but, as we argued earlier, we want to resist the presumption that working on understanding is limited to conceptual talk. What forms of talk then have we identified, and how have we specified and measured them, using teacher questions for the moment as a proxy measure of classroom talk?

In broad terms, in this particular project we have identified, specified and measured three kinds of classroom talk: conceptual talk, performative talk, and procedural talk. *Performative talk* is talk that focuses on the use of closed questions by the teacher to test student knowledge and/or understanding and the efforts of students to give the right answer (a performative demonstration of knowledge). Performative talk includes teacher statements or hints that attempt to get the student to get (or guess) the right answer (e.g. "If we multiply an even number by an even number shouldn't we get an even number as well...." "Have you considered...." etc.) Closed questions are questions that have a single right answer. Closed questions typically focus on propositional knowledge and are generally located in IRE sequences. More broadly, they are characteristic of pedagogies that focus on transmission and reproduction.

Procedural talk is talk that focuses on how students complete a process or task specific to a discipline, subject or area of study. It can refer to quite general procedural issues — methods of inquiry, particular methodologies, genres of work — or, more narrowly, to task-specific scripts, rules, procedures, strategies, algorithms, heuristics involved in solving a problem or doing knowledge work. Procedural questions are generally *how* rather than *why* or *what* questions. But there is good reason to believe that procedural fluency and conceptual understanding are interdependent. In Mathematics, for example, where there is generally a lot of procedural talk, procedural talk is often promoted by a question from a teacher to students ("What's the standard procedure, algorithm or rule for solving this kind of problem?" "Are there alternative procedures for solving this kind of problem?" "Are there better procedures for solving this problem?") or by a procedural question from a puzzled student to a teacher (e.g. "I don't understand how to solve this problem?" "What's the best way to solve this problem?"). Importantly, procedural talk is often closely associated with *conditional knowledge* — that is, knowledge of when to use specific procedures or algorithms. Thus, procedural talk often includes talk about *when* as well as *how*.

Conceptual talk is indicated by evidence of one or more forms of classroom exchange that focus on clarifying meaning, exploring how to solve problems, inviting alternative perspectives or solutions, making connections and establishing conceptual relationships, offering reasons and explanations, discussing epistemic standards, framing and reframing an issue, talking about learning, and potentially engaging in dialogical exchanges.

So, what then do students report about the nature of classroom talk in Singapore? Importantly, our survey questions asked students how often their teachers asked certain kinds of questions in initiating verbal exchanges in the class — it did not ask them to report on the full exchange between students and teachers. These questions broadly map onto the three afore-mentioned categories of talk. So, at best, we are only capturing a small part of the exchanges that teachers initiate, and nothing at all of the exchanges that students initiate with their teachers or with each other (if permitted).

Table 2 reports the findings from our survey of Secondary 3 Mathematics classes. Mean scores are rank ordered, on a frequency scale from 1 to 5. Even a casual glance suggests that teachers are more likely to ask performative and procedural questions rather than conceptual questions. A confirmatory factor analysis (Figure 1) supported our hypothesis of a tripartite model of teacher questions — performative, procedural and conceptual — but also indicated that we need to disaggregate conceptual questions into four subcategories: *clarifying* (e.g., "what do you mean?"), *connecting* (e.g., "what is the relationship between these two ideas?"), *explanatory* (e.g., "can you give me reasons for why you think that"?) and *epistemic* (e.g., "what makes this a reasonable statement?") (Figure 1). Happily, the model produced strong factor loadings and good fit statistics, indicating that the latent constructs (the classroom talk categories) substantially account for patterns of covariance between the various indicators.

Table 3 reports the mean scores and standard deviations of the three broad categories of classroom talk in Mathematics. In Mathematics, on a frequency scale from 1 to 5, the mean score for performative talk (3.67) outpaces the mean score for procedural talk (3.46). There is then a big gap to conceptual talk (3.18), indicating that teachers ask far fewer conceptual questions than they do performative and procedural ones. Among the subcategories of conceptual talk, the mean scores for both clarifying and connecting talk are 3.23, with explanatory talk on 3.15 and epistemic talk bringing up the rear at 3.11. This is not especially good pedagogical news, given the cognitive capabilities of understanding talk, as Barnes, Alexander, Nystrand et al. and Mercer and Littleton have emphasized, and the strategic importance of complex forms of communication in most models of 21st century skills.

Table 2

Classroom talk: Teacher questions secondary 3 Mathematics (2010)

	Secondary 3 Mathematics N = 1166		
	Mean (1-5)	SD	Talk Category
What is the correct answer to this problem?	3.69	.937	Performative
Is this answer right or wrong?	3.65	.971	Performative
How did you come to that answer?	3.52	.948	Procedural
What's the best way of solving this problem?	3.52	.948	Procedural
How do we know if this solution is correct	3.43	.956	Procedural
What is the next logical step to solve this problem	3.37	.991	Procedural
Could you explain what you mean?	3.31	.958	Clarifying
How would you know whether that is true or not?	3.28	.960	Explanatory
Can you give me reasons for why you think that?	3.27	.990	Explanatory
Is this statement true or false?	3.25	.982	Performative
What is the relationship between this mathematical idea and that one (e.g., speed and time)	3.23	.962	Connecting
What do you mean?	3.16	.975	Clarifying
What makes this formula true?	3.11	1.066	Epistemic
How good an explanation is that?	3.05	1.016	Explanatory
What makes this a reasonable guess?	3.03	.983	Explanatory

Table 3

Mean scores/SD: Revised classroom talk scales, Mathematics and English, 2010

	Mathematics		English		
	Mean (1-5)	SD	Mean (1-5)	SD	*Cohen'sd*
Performative Talk	3.67	.849	3.56	.877	*.127*
Procedural Talk	3.46	.774	3.30	.834	*.199*
Conceptual Talk	3.18	.743	3.33	.694	*.209*
-Clarifying Talk	3.23	.852	3.43	.858	*.234*
-Connecting Talk	3.23	.962	3.15	.809	*.090*
-Explanatory Talk	3.15	.804	3.51	.812	*.446*
-Epistemic Talk	3.11	1.066	3.24	.916	*.130*

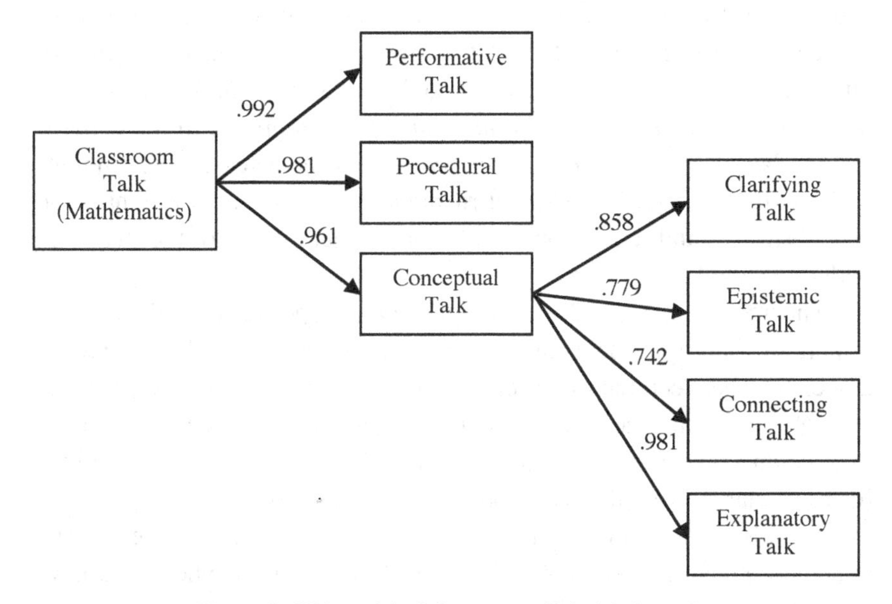

Figure 1. CFA model of classroom talk in Mathematics

Goodness-of-fit statistics	
Chi-Square / d*f* / *p*-value	331.186/ 74 / .000
CFI / TLI	.967 / .959
RMSEA (90% C.I.)	.055 (.049-.061)
SRMR	.043

The picture that this presents appears clear enough, but the correlations reported in Table 4 muddy the waters substantially and suggest quite a different kind of story. Specifically, the correlations suggest that there are very high levels of covariance between the three kinds of classroom talk, indicating that the relationships between the three forms of talk are complementary rather than orthogonal: teachers do not ask one kind of question and neglect others, but, up to a point, tend to ask all three kinds of questions. This is a very important finding, because it hints strongly that performative talk is not necessarily at odds, let alone at war, with other two main forms of classroom talk, and that the revisionist understanding of IRE might very well hold water. Of course, this finding applies only at a very general level, given that we asked students to report on the relative frequency of the different kinds of questions their teachers *generally* asked. *The survey data cannot show how tightly coupled these different forms of questions are at the activity, task or lesson level where it really matters*. Nonetheless, these findings underscore the importance of not rushing to judgement about what kind of questions (and classroom talk more generally) "do the work of understanding."

The mean scores and correlation matrix reported provide us with important findings about the relative importance and the degree of association between the three forms of classroom talk. But again they do not tell us about the specific causal pathways between them: How exactly and specifically are the different kinds of talk related? Which forms of classroom talk predict other kinds of classroom talk? To answer these kinds of questions, again at a very general level, we need to run structural equation models of the relationships. Happily, when we do, we begin to get a rich causal picture of the intricate set of relationships between performative talk, on the one hand, and procedural and conceptual talk, on the other, in ways that belies the generally negative press IREs have had in recent decades.

Table 4
Correlation matrix: Latent constructs secondary 3 Mathematics (and English)

	Performative Talk	Procedural Talk	Conceptual Talk	Clarifying Talk	Connecting Talk		Explanatory Talk	Epistemic Talk
Performative Talk	1							
Procedural Talk	673** (.447**)	1						
Conceptual Talk	.538** (.539**)	.773** (.738**)	1					
Clarifying Talk	.477** (.508**)	.605** (.503**)	.754** (.797**)	1				
Connecting Talk	.413** (.428**)	.584** (.674**)	.779** (.830**)	.411** (.525**)	1			
Explanatory Talk	.501** (.469**)	.747** (.659**)	.871** (.828**)	.632** (.588**)	.587** (.599**)		1	
Epistemic Talk	.367** (.364**)	.580** (.584**)	.823** (.818**)	.455** (.493**)	.498** (.609**)		.638** (.542**)	1

*** Significant at .001 level*

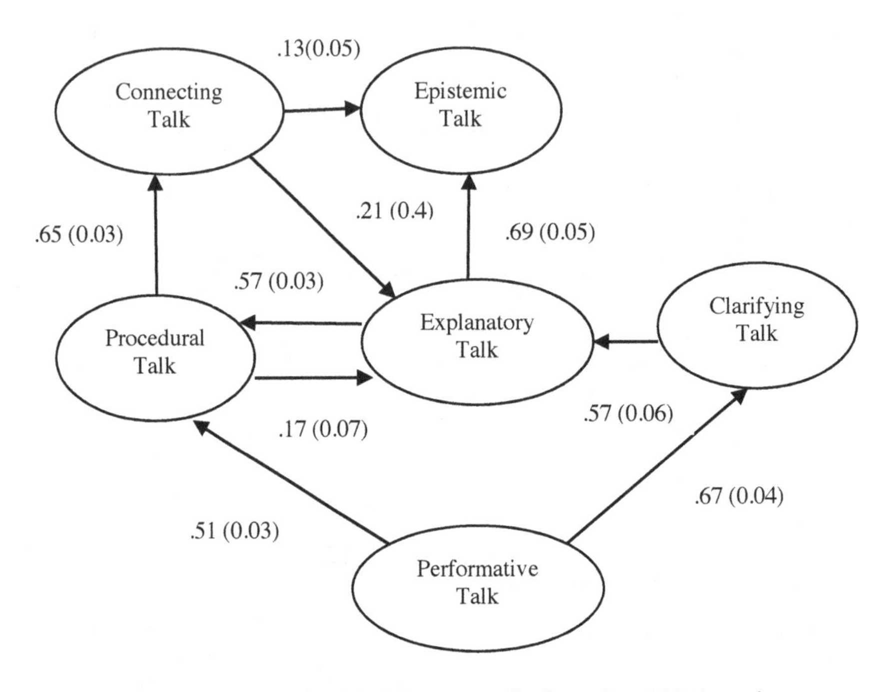

Figure 2. SEM model of classroom talk: Secondary 3 Mathematics

Goodness-of-fit statistics	
Chi-Square / df / p-value	7.096 /6 /.3145
CFI / TLI	1.00/.999
RMSEA (90% CI)	.012(.000-.041)
SRMR	.008

Figure 2 represents a SEM model of classroom talk with all the categories for conceptual talk included. Several features of the model are particularly noteworthy. First of all, it is a very strong model with exceptionally good fit statistics on all relevant criteria. Second, the model clearly identifies performative talk as the key driver of the discursive regime in Secondary 3 mathematics classes. This is perhaps not surprising given the mean scores for performative talk reported in Table 3. But the story does not end there, since the pathways identified in Figure 2 indicate a strongly integrated and coherent discursive regime in

Secondary 3 mathematics that belies the impression of a discursive system preoccupied with performative recitations or IRE sequences. This is because there are two very strong pathways from performative talk, on the one hand, to procedural (.51) and clarifying talk (.67), on the other. The first indicates that teachers very often follow up a performative question with a procedural question ("What is the next logical step to solve this problem") while the second indicates that teachers often respond to student responses to their performative questions with a clarifying question ("what do you mean"?). In effect, to put it another way, performative questions strongly predict procedural and performative questions. Critically, teachers very often follow up their clarifying questions with explanatory questions (.57) ("How did you get that answer?") which in turn is followed up by either a procedural question (.57) or, even more often, an epistemic question (.69) ("What makes this formula true?"). But beyond these pathways, there is a strong pathway from procedural talk to a key form of conceptual talk — connecting talk (.65) ("What is the relationship between these two ideas?" "What are the implications of your argument?") — that is critical to building conceptual networks at the cognitive level that help develop understanding and knowledge building. From connecting talk there are weaker pathways to epistemic talk (.13) and to explanatory talk (.21), followed up by a very strong pathway from explanatory talk to epistemic talk (.69), as we noted earlier.

The SEM model also indicates that explanatory talk has an important non-recursive, reciprocal relationship with procedural talk in which explanatory talk not only strongly predicts procedural talk (.57) but procedural talk, in turn, predicts explanatory talk, although not nearly as strongly (.17). This suggests that teachers use explanatory talk and procedural talk interchangeably, but the stronger path from explanatory talk suggests a prioritization of procedural talk in the classroom, in the Singaporean fashion. Nevertheless, this reciprocal relationship brings about a strong (in fact, the strongest estimate in the model) path from explanatory talk to epistemic talk. The combined indirect paths from procedural talk to epistemic talk are statistically significant at .352. Importantly, the non-recursive path also allows another set of indirect paths to be calculated from explanatory talk, through procedural talk, connecting talk to epistemic talk, resulting in a statistically significant estimate of .200. The total effect of explanatory talk on epistemic talk is therefore a very substantial .89 (.69 + .20). In both pathways from performative talk to epistemic talk, explanatory talk is the key moderating variable between performative talk and epistemic talk. Procedural talk and clarifying talk are each moderator variables for the path from performative talk to explanatory talk i.e., the direct path between performative talk and explanatory talk is stronger than each of the two paths performative-procedural-connecting-explanatory (.984 x .620 x .182 =.111) and performative-clarifying-explanatory (.802 x .365 = .293). In addition, explanatory talk is the mediator variable for the path from connecting talk to epistemic talk. This means that in order to do epistemic talk, explanatory talk is an important intervening variable as far as connecting talk is concerned.

From a disciplinarity perspective, the strength of this relationship is a hugely important — and, from the Ministry's perspective, reassuring — finding, since it tells us that teachers broadly recognize that understanding the epistemic norms governing mathematics depends on prior understanding of the reasons why a solution works or does not work and is not simply an isolated form of knowledge that has no clear links to procedural or other forms of conceptual knowledge. More broadly, it indicates that mathematics teachers recognize (even if they would not quite describe it in quite this terms) that epistemic talk represents the apex of the hierarchy of classroom talk because it asks deeper questions of what makes something true and not just why it is true. Indeed, it is our view that improving the intellectual quality of classroom discourse — and task design — will depend significantly on an enhanced epistemic presence in mathematics classrooms. As the distinguished classroom discourse analyst Douglas Barnes (2008, p.14) recently wrote in a semi-valedictory paper, over the decades of researching and teaching about classroom talk, he has come to the conclusion that the teachers need to focus far more than they do on epistemic issues in their teaching:

> As the years have passed I became more and more convinced that learners should be given more access to the grounds upon which the

knowledge they were learning was based. Knowledge is too often presented as if it is beyond challenge and beyond the examination of alternatives. Tasks are often set without teachers indicating in terms that learners can understand what criteria will be used to judge success or failure. I am convinced not only that conscious participation in learning is essential, but also that in order to achieve active learning learners should be encouraged to be reflective and critical.

Beyond this, the strength of the pathways from performative talk to procedural talk and conceptual talk (either directly by way of clarifying talk or indirectly by way of procedural talk and from their connecting talk or explanatory talk), and even more challengingly, from procedural talk to epistemic talk, underscores our contention, hinted at earlier, and explicitly supported in general terms by Wells, Mercer and others, that performative talk is not necessarily *just* performative talk, but often closely associated with cognitively complex forms of interaction and "working on understanding." It all depends on the classroom situation or context, the purposes and understandings of the teacher, the nature of the tasks students are engaged in, and whether or not teachers decide to follow up a performative question with a procedural or conceptual question. Indeed, although our data is highly decontextualized, the student survey data offers a strongly revisionist picture of the performative talk.

Still, despite the (apparent) virtues of the mathematics discursive regime in Secondary 3 mathematics classes, the structure of teacher questioning could be improved in at least two respects. The path, for example, from connecting talk to epistemic talk is much weaker (.13) than it ought to be on normative grounds, given our contemporary research understanding of the role of conceptual knowledge in building a coherent and integrated understanding of a disciplinary field. By the same token, the path from connecting talk to explanatory talk (.21) is weaker than it might be. Finally, we are a little surprised by the relatively weak pathway from procedural knowledge to explanatory knowledge (.17) — we had hoped for a stronger pathway, reflecting our understanding of the important contribution that procedural knowledge can play in developing conceptual knowledge, as we argued in our earlier chapter. Going forward, we recommend that both in-service and pre-service programs in Singapore attend to strengthening these pathways.

4 Conclusion

While, performative questions clearly initiate the discursive regime in Secondary 3 Mathematics, performative questions are not necessarily hostile to "working on understanding," at least as measured by student reports of the kind of questions that teachers ask them in class. This doesn't necessarily mean that students in fact do develop understanding — we won't know the answer to that question until next year after we have finished our classroom observation data — but teachers clearly have a reasonably developed, if implicit, causal understanding of important relationships between the different kinds of questions that they ask. In particular, performative questions, as a leading proxy for performative talk, have strong direct and indirect relationships to the two key pivots of the discursive regime in Singaporean mathematics classroom — procedural and explanatory talk. This is an important finding and we think helps explain the success of Singaporean students in international assessments although overall levels of classroom talk are relatively low. It also suggests that we need to be extremely cautious in judging "closed" teacher questions and performative talk, including IRE, as a hum-drum ritualized exchange that has little or no capacity to prompt students to work on understanding. This is *not* to say that teachers could not ask conceptual questions more often than they do, or that they could *not* strengthen the relationships between key forms of talk that are central to developing conceptual understanding, because clearly they could. In particular, we believe that the quality of classroom talk and student learning could be improved by forging tighter relationships between connecting talk, on one hand, and explanatory and epistemic talk, on the other, both of which are fundamental to "working on understanding."

Acknowledgement

The authors want to acknowledge the contribution of the following NIE staff involved in the development of the data base used in this study: Serena Luo, Sheng Yee Zher, Khin Maung Aye, Tan Teck Kiang and Loo Siok Chen.

References

Alexander, R. (1992). *Policy and practice*. London: Routledge.

Alexander, R. (2008a). Culture, dialogue and learning: Notes on an emerging pedagogy. In N. Mercer, & S. Hodgkinson (Eds.), *Exploring talk in schools* (pp. 91-114). London: Sage.

Alexander, R. (2008b). *Essays on pedagogy*. London: Routledge.

Alexander, R. (2008c). *Towards dialogic teaching: Rethinking classroom talk* (4thed.). York: Dialogos.

Alexander, R. (2010). *Dialogic teaching essentials*. Singapore: National Institute of Education.

Barnes, D., Britton, J. & Torbe, M. (1986). *Language, the learner and the school*. London: Penguin.

Barnes, D. (1992). *From communication to curriculum* (2nded.). Portsmouth, NH: Boynton/Cook-Heinemann.

Barnes, D. (2008). Exploratory talk for learning. In N. Mercer, & S. Hodgkinson (Eds.), *Exploring talk in school* (pp. 1-16). London: Sage.

Brown, G. & Wragg, E.C. (1993). *Questioning*. London: Routledge.

Cazden, C. B. (2001). *Classroom discourse: The language of teaching and learning*. 2nd edition. Portsmouth, NH: Heinemann.

Dillon, J. (1994). *Using discussion in classrooms*. Buckingham: Open University Press.

Doyle. P. & Hong, H. (2009). *Compiling a multimodal corpus of educational discourse in Singaporean schools*. Singapore: CRPP/OER, 2009.

Francis, P. (2002). Get on with your talk. *Secondary English Magazine, 5*(4), 28-30.

Galton, M. (2009). Group work: Still a neglected art? *Cambridge Journal of Education, 39*(1), 1-6.

Galton, M. J., Croll, P., & Simon, B. (1980). *Inside the primary classroom*. London: Routledge & Kegan Paul.

Galton, M., Hargreaves, L., Comber, C., Wall, D. & Pell, A. (1999). *Inside the primary classroom: 20 years on*. London: Routledge.

Hardman, F. (2008). Teachers' use of feedback in whole-class and group-based talk. In N. Mercer & S. Hodgkinson (Eds.), *Exploring talk in school* (pp. 131-150). London: Routledge.

Hattie, J. (2009). *Visible learning: A synthesis of over 800 meta-Analyses relating to achievement*. London: Routledge.

Heap, J. L. (1985). Discourse in the production of classroom knowledge: Reading lessons. *Curriculum Inquiry, 15*(3), 249-279.

Hodgkinson, S., & Mercer, N. (2008). Introduction. In N. Mercer & S. Hodgkinson (Eds.), *Exploring talk in school*. London: Sage.

Hogan, D., Chan, M., Ridzuan Abd Rahim, Luo, W., Sheng, Y. Z., Khin, M. A., Towndrow, P., & Kwek, D. (2012). Policy and practice: Teach less learn more and instructional methods in Secondary 3 English and mathematics in Singapore, 2010. In Z. Deng, S. Gopinathan & C. Lee (Eds.), *Globalization and the Singapore curriculum: From policy to classroom*. Singapore: Springer.

Hogan, D., Towndrow, P., Abdul Rahim, R., Chan, M., Luo, S., Sheng, Y. et al. (2011). *Interim report on pedagogical practices in Singapore in Secondary 3 mathematics and English, 2004 and 2010* National Institute of Education, Singapore.

Lefstein, A., & Snell, J. (2011). Classroom discourse: The promise and complexity of dialogic practice. In S. Ellis, E. McCartney & J. Bourne (Eds.), *Insight and impact: Applied linguistics and the primary school* (pp. 165-185). Cambridge: Cambridge University Press.

Lemke, J. L. (1989). *Using language in the classroom*. Oxford: Oxford University Press.

Lemke, J. L. (1990). *Talking science: Language, learning and values*. Norwood, NJ: Ablex.

Mehan, H. (1979). *Learning lessons: Social organization in the classroom*. Cambridge, MA: Harvard University Press.

Mercer, N. (1992). *Talk for teaching and learning*. In K. Norman (Ed.), *Thinking voices: The work of the National Oracy Project* (pp. 215-223). London, UK: Hodder & Stoughton (for the National Curriculum Council).

Mercer, N., & Littleton, K. (2007). *Dialogue and the development of children's thinking: A sociocultural approach*. Oxon: Routledge.

Myhill, D. & Dunkin, F. (2005). Questioning learning. *Language and Education, 19*(5), 415-428.

Nassaji, H., & Wells, G. (2000). What's the use of 'triadic dialogue'?: An investigation of teacher-student interaction. *Applied Linguistics, 21*, 376-406.

Nystrand, M. (1997). *Open dialogue: Understanding the dynamics of language and learning in English classrooms*. New York: Teachers College Press.

Nystrand, M., & Gamoran, A. (1991). Instructional discourse, student engagement, and literature achievement. *Research in the Teaching of English, 25*, 261-290.

Nystrand, M., Wu, L. L., Gamoran, A., Zeiser, S., & Long, D. A. (2003). Questions in time: Investigating the structure and dynamics of unfolding classroom discourse. *Discourse Processes, 35*(2), 135-198.

Postman, N. & Weingartner, C. (1969). *Teaching as a subversive activity*. New York: Delta.

Sinclair, J. M., & Coulthard, M. (1975). *Towards an analysis of discourse: The English used by teachers and pupils*. London: Oxford University Press.

Smith, F., Hardman, F., Wall, K., & Mroz, M. (2004). Interactive whole class teaching in the National Literacy and Numeracy Strategies. *British Educational Research Journal, 30*(3), 395-411.

Wells, G. (1999). *Dialogic inquiry: Towards a sociocultural practice and theory of education*. Cambridge: Cambridge University Press.

Wells, G., & Arauz, R. M. (2006). Dialogue in the classroom. *The Journal of the Learning Sciences, 15*(3), 379-428.

Wolfe, S. & Alexander, R. (2008). *Argumentation and dialogic teaching: Alternative pedagogies for a changing world*. Retrieved December 1, 2011, from http://www.beyondcurrenthorizons.org.uk/argumentation-and-dialogic-teaching-alternative-pedagogies-for-a-changing-world/

Chapter 10

Mathematics Classroom Discourse Through Analogical Reasoning

Kyeong-Hwa LEE

Analogical reasoning is recognized as an efficient means of problem solving. However, few research studies have examined how to integrate analogical reasoning into mathematics lessons and study the relationship between use of analogical reasoning and discourse in mathematics classroom. This study presents an analysis of discourse produced during a mathematics lesson that involved analogical reasoning. Observation analysis of the class and students' final constructs supported the notion that analogical reasoning facilitates rich discourse, interactions between the class teacher and the students themselves.

1 Introduction

Many studies have shown analogical reasoning to be a useful form of mathematical reasoning for constructing mathematical knowledge by extensively utilizing pre-existing knowledge (English & Sharry, 1996; Lakatos, 1976; Lee & Sriraman, 2011; Polya, 1954; Sriraman, 2006). Analogical reasoning is a kind of conceptual strategy to make conjectures about novel phenomena (Richland, Morrison, & Holyoak, 2006) and to transfer pre-existing knowledge to new contexts (Lakatos, 1976; Polya, 1954; Lee & Sriraman, 2011). Analogical reasoning has been explained by the ability to perceive and operate on the basis of corresponding relational or structural similarity in objects (English, 2004; Richland, Holyoak, & Stigler, 2004). Despite many research studies on the nature of analogical reasoning as a form of mathematical thinking, little is known about analogical reasoning in mathematics lessons as Richland and his colleagues pointed out (2004).

There is a need to carefully link the role of communication in mathematical reasoning (Anthony & Hunter, 2010; Cobb, Boufi, McClain, & Whitenack, 1997; Lampert, 1990; Sfard, 2008; Walshaw & Anthony, 2008). Regrettably, several studies contend that Korean students learn mainly from systematic teacher presentations and drills and practices rather than through verbal interaction in the mathematics class (Park, 2001 & 2004; Lee, 2010). To address this issue and improve classroom interaction, mathematical communication skills have been emphasized since the 1997 national mathematics curriculum. Moreover, since 2000, all secondary school textbooks include tasks that offer opportunities for mathematics communication. Regardless, improving verbal interaction in mathematics classes is still a difficult task in Korean mathematics education. A recent study by Lee and Sriraman (2011) highlights analogical reasoning's benefits when directly incorporated into the mathematics learning process. These researchers using a re-conceptualized classical analogy termed Open Classical Analogy (hereinafter referred to as OCA) reported positive learning results from OCA implementation. The purpose of this chapter is to present analysis of mathematics class discourse produced during a lesson centred on an OCA-type problem suggested by Lee and Sriraman (2011).

2 Analogical Reasoning and Discourse

In its simplest form analogical reasoning can be defined as reasoning that occurs when pursuing or making similarities among objects or attributes. It is sometimes referred to as analogy in related and relevant literature. English (1997; 2004) noted that analogical reasoning takes various forms in mathematics teaching and learning. One such form called *Classical Analogy* refers to reasoning that takes the form A:B::C:D, where the C and D term relationship corresponds to the A and B term relation (English, 2004). Lee and Sriraman (2011) re-conceptualized *Classical Analogy* and termed it *Open Classical Analogy*. They defined their new type of analogy as:

> An analogy that requires learners to authentically search for the "B", "C", and "D" terms of a classical analogy (p. 126).

The requirement for learners to look for different combinations of "B", "C", and "D" terms; that is, the three types of OCA problems, can be regulated through the provision or absence of terms given by teachers. For instance, educators may opt to provide learners with the "A" and "C" terms of a classical analogy and ask students to search for the "B" and "D" terms as in the following problem:

> Knowing that a tetrahedron is similar to a triangle, conjecture a tetrahedron property analogous to a property of a triangle that you know. Explain your answer.

Likewise, learners could be asked by educators to solve problems that provide the terms "A" and "B" and require the seeking of the "C" and "D" terms:

> The interior angles of a triangle sum to two right angles. Conjecture a similar property for any geometric figure that is analogous to a triangle. Explain your answer.

The third type of OCA problem is to provide only the "A" term:

> Select a geometric figure that is analogous to a triangle and conjecture a property that is analogous to a property of a triangle. Explain your answer.

Drawing on these three forms it is possible to modify many of the problems given or suggested in school mathematics curricula to create OCA type problems. For example, the following problem presented in Cañadas, Deulofeu, Figueiras, Reid, and Yevdokimov (2007) can be rephrased to an OCA type problem in which only the "A" and "B" terms are given.

[Original problem]
Given a triangle ABC and a point P inside the triangle construct the three lines from each vertex A, B, C to the point P. What can you say about the relationships between the lines and the sides of the triangle? (p. 59)

[Rephrased problem]
Three median lines of a triangle meet at a single point. Select another line, other than a median line, and conjecture a property that is analogous to the given property. Explain your answer.

According to Sternberg (1977), reasoning with *Classical Analogy* can be analyzed in terms of components of information processing involved: *identifying* each term of the analogy by encoding the attributes; *inferring* the relationship between the terms within each pairing is determined; *mapping* by linking the A:B pairing to the C:D pairing by searching a bridging inference on their relational similarity; and *applying* by generating a suitable term to complete the analogy. It is reasonable to extend this componential approach to reasoning with OCA because all the components involved in *Classical Analogy* are to be controlled to solve OCA type problem.

A number of research studies have demonstrated that classroom discourse can play a significant role in supporting students' mathematics learning (Cobb, Boufi, McClain, & Whitenack, 1997; Lampert, 1990; Sfard, 2008). My purpose in this chapter is to suggest possible relationships between classroom discourse and analogical reasoning by the students who participate in. To this end, I focus on particular discourses that are connected to components of reasoning with OCA.

For the most open OCA type problem in which only the "A" term is given, discourse throughout the solving process between the class teacher and the class as a whole as well as among students themselves involve five specific cognitive activities: (1) problem understanding by recognizing the structure of analogy; (2) recollection of base object attributes, i.e. *identifying* activity; (3) selection of a choice target after *inferring and mapping* the relation of A:B pairing; (4) analogy consolidation by *applying* the relation of the A:B pairing to the one of the C:D pairing; and (5) reflection on the entire analogy generation process. More specifically, in the aforementioned OCA problem solving process, discourse during the "problem understanding" phase encompasses recognition of structural aspects of the problem; i.e., students need to verbalize both what is given and what need seeking (see Discourse 1, Figure 1). Discourse for "recollection of base object attributes" should centre on identifying conceptual aspects of base object (see Discourse 2, Figure 1) whereas discourse during the "selection of a choice target" phase of the solving process needs to involve language to infer and map specific relation in the A:B pairing (see Discourse 3, Figure 1). At the next stage, discourse for "analogy consolidation" should embody arguments for finding mathematical support for the declared similarity between base and target objects by applying the mapped relation to the C:D pairing (see Discourse 4, Figure 1). Finally, discourse produced during the phase "reflection on the entire analogy generation process" consists of speculations at each step and declaration of the final analogy in terms of its value and meaning (see Discourse 5, Figure 1).

In presenting observations of a mathematics lesson that implemented an OCA problem solving, I make a distinction between teacher-initiated discourse from student-initiated to clarify teacher' and students' contribution to the moulding of discourse and consider possible relationships between individual OCA process and the classroom discourse. Finally, I characterize the discourse and conclude by summarizing its significance while integrating OCA in mathematics lessons.

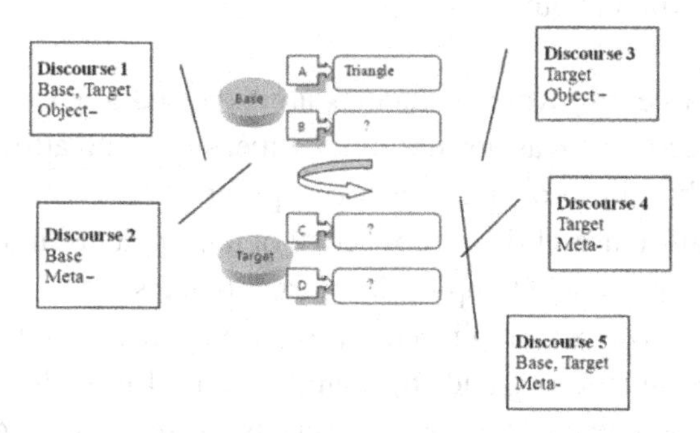

Figure 1. Discourses focused on the components of reasoning with OCA

3 Context of Classroom Observation

Participants in this study included Korean 8th grade (Year 8) mathematics class. The class teacher has worked at the middle school level for 10 years and volunteered to participate in the research project because of a personal drive to learn from the experience. She expressed genuine interest in creativity development through the activation of student knowledge construction—and was eager to build a discourse-rich mathematics classroom and mould discourse on knowledge creation. The teacher had extensive reading on analogy use in mathematics including this researcher's prior research studies and was familiar with OCA type problem sets.

Interview and observations data were collected across three consecutive 45-minute lessons. The teacher partook in three one-hour-long interviews before and after each of the three lessons. As part of the classroom observation, video-based data was collected on discourse relevant to analogical reasoning as students engaged in solving OCA problems.

Of the three types of OCA problems, the teacher opted for the most open one; the problem type that provides few known variables. Below is an example of the problem presented to her students.

> Select a geometric figure that is analogous to a triangle and conjecture a property of that figure that is analogous to a property of a triangle. Explain your answer.

By grade 8 (Year 8) Korean students are expected to know a number of triangle aspects such as its definition, measurement attributes, and properties in the form of mathematical propositions. Students were encouraged to make use of their knowledge of a triangle to find relational similarities in order to select appropriate target objects.

Data analysis was accomplished via two methods: (a) review of the observed discourse meaning and (b) identification of how discourse was initiated and maintained. Lesson transcripts were coded to identify critical events and episodes that revealed significant features of the relationships between analogical reasoning and discourse in mathematics classes.

4 Findings

Analogical thinking demands that students create similar knowledge by eliciting similarities and differences or recalling pre-existing knowledge (English, 2004; Leech, Mareschal, & Cooper, 2008). In other words, analogical thinking necessitates separation, connection, transition, and dialectical integration between previous and present learning experiences. This study investigates how these various analogical thinking functions influence the form and content of classroom discourse. Analysis and presentations of the discourse that emerged is by two types: teacher-initiated and student initiated.

4.1 *Teacher-initiated discourse*

Initial observation of the classroom discourse suggested that enactment of the OCA-type tasks supported active involvement in classroom communication. However, deeper analysis of that discourse revealed that the teacher played an essential role in the initiation of and maintenance of discourse by asking questions. In particular, discourses for the first three cognitive activities (Discourse 1, 2, & 3) involved were mainly initiated and maintained by the class teacher with the whole class. In contrast to these discourses, discourse for analogy consolidation (Discourse 4) and reflection of analogy generation process (Discourse 5) were initiated not only by the teacher to the whole class but also by students to the members during group work.

Example 1: Teacher Initiated Discourse to promote understanding of an OCA-type task

The OCA-type task used in this study was unfamiliar to students. Most students did not fully comprehend the structure of the problem and, as a result, immediately set about on a search for a similar object. The similar object first proposed by students was a superficial target object; it shared only a perceptual part of an attribute with the base object. To scaffold the students the teacher initiated discussion that would lead students to grasp the superficiality of their target object choice as follows:

[T721] T: Now, it's time to share your conjectures. Let's enjoy looking at classmates' analogies for a triangle. S1, would you please share your proposed figure?
[T722] S1: (pointing to the figure he drew) This figure is the answer.
[T723] T: Would you please explain your analogy?
[T724] S1: It is this figure.
[T725] T: Your analogy is only this figure?
[T726] S1: Because it is similar to a triangle
[T727] T: Ah! Why do you think they are similar?
[T728] S1: Don't you think they are similar?
[T729] T: Perhaps. But what is your reason for saying they are similar?
[T730] S1. Oh! I forgot. I forgot!
[T731] T: Pardon?
[T732] S1: There are no B and D terms. I need the B and D parts to write an analogy.

Solving this OCA-type task required students to find a target object, alongside consideration of common or similar properties between a potential target object and the base object. However, a number of students bypassed the search for a target figure that has perceptual or surface similarity to look at objects of common geometric or relational similarity. The Teacher Initiated Discourse (TID) in the above example helped students recognize the importance of relational similarity and directed students' focus away from making purely superficial connections towards looking for structural connections between their existing knowledge and new mathematical knowledge (lines T727&T729).

As expected, discourse that facilitated problem understanding (Discourse 1) was frequently observed between the teacher and the whole class and amongst students during seat work. However, despite the high occurrence of Discourse 1, several students were confused about task goal in a similar way to S1 in the above episode and thus were likely to benefit from the explicit teacher redirection back to the first phase of the solving process—problem understanding (lines T729-T732).

Example 2: Teacher Initiated Dialogue to evaluate initial solutions to the OCA-type task

Based on Discourse Criteria 1 to 5 (see Figure 1), the teacher mainly conversed with the class as a whole and on occasion; with an individual student when deemed necessary. Because there are generally multiple solutions for OCA-type tasks, having the opportunity to share solutions and understand rationale behind the proposed solutions is essential. The teacher facilitated solution sharing and evaluation of their analogies. In other words, she guided students in a discussion about initial targets and their respective properties. These discussions personify the thought process toward discovery of new objects and properties using meta-analysis of existing knowledge. The following teacher feedback illustrates how the teacher reaffirmed norms of sharing and evaluation of their analogies.

[T908] T: Wow, you have so many various solutions. I never expected this. I thought you'd come up with only one or two solutions. You worked very hard and got many answers. Great job. How about showing the rest of the class and having other students learn from your work? S2, would you mind presenting your group's solutions?

[T990] T: Look. All three solutions are very similar, yet all three are different. Do you see that? Let's discuss the value of each of these target objects. Like mathematicians, let's enjoy what we have found, I mean, your target object choices. What do you think of S2's target object? Is there anything you like about it? How should we assess its value? Any suggestions?

Students began to recognize that the mathematical meaning of target object differ when knowledge utilized in the search for a target object is superficial compared to when it is structural. In other words, students concluded that analogy generation based on simple knowledge of a triangle's definition such as "having three sides" does not produce meaningful results. In contrast, they deemed that analogy generation based on advanced knowledge of a triangle such as "the center of gravity" or "the circum center" can lead to grounded results. Moreover, during the process of analogy generation, students came to realized that the discovery of new triangle facts could also produce good target objects. Hence, by examining value, students could identify whether it was important to use pre-existing knowledge or to reconstruct pre-existing knowledge by incorporating new perspectives. Transcribed dialogues from student group in the class illustrating this realization process are presented in the next sections.

4.2 *Student-initiated discourse*

Although the teacher often initiated the classroom discourse, at times discourse was started voluntarily by the students themselves. Importantly, this study revealed that students actively participated in discourse related mathematical knowledge creation through genuine analogy production. This discourse encompassed meta-analysis of mathematical concepts and/or properties.

Example 3: Student Initiated Discourse in the process of comparing target objects

This study's focus OCA-type task can be best solved by finding a target object and its attribute and properties simultaneously. However, at the onset of the solution process, most students focused their attention towards finding a target object. It was only after proposing a target object that they tried to conform whether the target object choice was a viable structure solution for the problem. The following dialogue between two students, S7 and S9, demonstrates this thought process.

[S107] S7: Is that your answer?
[S108] S9: It is different from mine.
[S109] S7: The teacher said there could be many answers.
[S110] S9: I know. They can be different, but shouldn't the target objects be based on the same reason?

The two students then individually reexamined their proposed target objects. By making comparisons to a triangle, they continued with their talk, offering, as justification, target object similarity to a triangle.

[S111] S7: I found this target using the fact that a triangle is a figure with a vertex, sides, and a face.
[S112] S9: I looked at a triangle as a geometric figure, but I focused on the sum of all interior angles of a triangle always equaling180 degrees.
[S113] S7: Are the reasons for our answers different?

This dialogue between S7 and S9 illustrates how students are able to reflect on the problem solving process and justify products of an initial analogy. The conversations show how two students can spontaneously engage in meaningful mathematical communication.

Example 4: Student Initiated Discourse in the process of evaluating target objects

The target objects suggested by students included both superficially similar objects and structurally similar objects. Irrespective of the type or number of targets offered, expectations that students share their target objects and discuss reasons for target and base object similarity, frequently generated opportunities for rich mathematical communication. As illustrated in the episode below, it was during this reflection process that students, S11 and S12, moved to target objects that were "new" mathematically.

[S507] S11: It is too obvious to just say a polygon is similar to a triangle.

[S508] S12: Well, their words are similar. Both words end with "gakhyung" ("gakhyung" is a Korean suffix meaning a figure with angles)

[S509] S11: Do you think we can find something new?

[S510] S12: Yes, it would probably be more meaningful to find a property similar to a triangle in a tetrahedron. It could probably be developed mathematically.

It is rare for students to have the above opportunity to engage in meta-analysis about mathematics when they are learning (lines [S508] & [S509]). Although accurate criteria definition and mathematical justification may be beyond students' ability, this kind of self-learning is a valuable new experience for students as discussed in Lee and Sriraman (2011).

5 Conclusion

This study investigated the nature and role of classroom discourse generated while students engage in an OCA task. In line with recent research studies on the discursive approach to mathematics learning (Kieran, Forman, & Sfard, 2002; Sfard, 2001 & 2008), this study showed how mathematical thinking was nurtured and expanded through mathematical communication. At the beginning of this study, analogical reasoning through OCA-type task solving was hypothesized to promote fruitful discourse in mathematics classes. Analysis of the OCA task enactment found that to achieve productive discourse, the class teacher needed to initiate talk at each phase of the solving process by directing students' attention towards analogical reasoning. Noticeably, despite major discourses were initiated by teachers, they were not only object-level but also meta-level that Sfard (2001) made a distinction; the former shows explicit intent whereas the latter resides in forms of utterances and mechanisms of interaction rather than in explicit content.

Example 1 is a part of a teacher initiated discourse in which the class teacher instructed students on the need to first comprehend the structural aspect of analogical reasoning. Through this teacher initiated discourse, the students recognized the necessity of seeing the problem structure. Noteworthy, the teacher did not directly point out ideas overlooked students as they solve the problem but, instead, led students to reflect on their problem solving process and clarify problem requirement. Example 1 depicted how the teacher guided students towards meta-level discourse.

In Example 2, the class teacher is observed inviting students share their various target objects and then channeling students to evaluate the validity and the meaning of those target objects. To inspire all students to participate in this evaluation, the teacher addressed the class as a whole, not individuals. This discourse generated various evaluation criteria from individual students. The criteria were based on their solutions and from discussions that took place within their student groups. Therefore, the entire analogy solution generation process including corresponding reflection on or evaluation of conjectured analogies could be individualized promoted by this meta-level discourse.

Example 3 and 4 specifically depicted how the teacher initiated discourses were reproduced as student initiated discourses. Essentially Example 3 reveals that students not only became aware of the need to focus on both the base and target object but also on the need for meta-level discourses of their activities and the constructs themselves, which was handled in Example 1. Students were observed spontaneously engaging in discourse on target object choices and the actions needed once satisfied with target object choices. Doubts and fears were often manifested in their conversations, but these doubts and fears were likely due to first-time exposure to an OCA-type task. While engaging analogy evaluation, Example 4 demonstrated students' use of discourse to share and analyze ideas. As aforementioned, the target objects and meta-level discourses of these target objects' properties led the students to i) appraise their prerequisite learning about figures such as knowledge of a triangle and ii) analyze newly developed conjectures stimulated by target object choice.

Opportunities for insightful analysis of prerequisite learning content and investigation of knowledge from various angles have been emphasized as essential channels for boosting mathematics learning. Irrespective of its simplicity, the result depicted in this chapter clearly shows that students can cultivate their thinking process by analyzing a simple object like a triangle from various angles and differentiate between what is known—pre-existing knowledge—and what needs to be learnt. This positive result, however, seems to rest on the class teacher's instructional strategy, which generate teacher initiated discourse that leads to student initiated discourse. In brief, this study suggests that effectual discourse initiated by the teacher leads students to understand an OCA-type task's goal and structure, thoughtfully analyze the analogy generation process and resultant proposed objects, and seek meaning in their resultant solutions by sharing and comparing with others in the class. Nevertheless, similar studies of students at differing school levels or that implement a variety of analogy tasks in different mathematical content areas would lead to better understanding of the function and role of discourse generated in mathematical classroom applying analogical reasoning.

Acknowledgement

The author is grateful to Glenda Anthony for her helpful comments on a previous draft.

References

Anthony, G. & Hunter, R. (2010). Communities of mathematical inquiry to support engagement in rich tasks. In B. Kaur & J. Dindyal (Eds.), *Mathematical applications and modelling: Yearbook 2010 Association of Mathematics Educators* (pp. 21-39). London: World Scientific.

Cañadas, M. C., Deulofeu, J., Figueiras, L., Reid, D., & Yevdokimov, A. (2007). The conjecturing process: Perspectives in theory and implications in practice. *Journal of Teaching and Learning, 5*(1), 55-72.

Cobb, P., & Boufi, A., McClain, K., & Whitenack, J. (1997). Reflective discourse and collective reflection. *Journal for Research in Mathematics Education*, *28*(3), 258-277.

English, L. D. (Ed.). (1997). *Mathematical reasoning: Analogies, metaphors, and images*. Mahwah, NJ: Lawrence Erlbaum Associates.

English, L. D. (Ed.). (2004*). Mathematical and analogical reasoning of young learners.* Mahwah, NJ: Lawrence Erlbaum Associates.

English, L. D. & Sharry, P. V. (1996). Analogical reasoning and the development of algebraic abstraction. *Educational Studies in Mathematics, 30,* 135-157.

Kieran, C., Forman, E., & Sfard, A. (Eds.). (2002). *Learning discourse: Discursive approaches to research in mathematics education*. Dordrecht, The Netherlands: Kluwer Academic.

Lakatos, I. (1976). *Proofs and refutations: The logic of mathematical discovery*. New York: Cambridge University Press.

Lampert, M. (1990). When the problem is not the question and the solution is not the answer: Mathematical knowing and teaching. *American Educational Research Journal*, *27*(1), 29-64.

Lee, K. H. (2010). Searching for Korean perspective on mathematics education through discussion on mathematical modeling. *Research in Mathematics Education*, *20*(3), 221-239. (In Korean)

Lee, K. H. & Sriraman, B. (2011). Conjecturing via reconceived classical analogy. *Educational Studies in Mathematics*, *76*(2), 123-140.

Leech, R., Mareschal, D., & Cooper, R. P. (2008). Analogy as relational priming: a developmental and computational perspective on the origins of a complex cognitive skill. *Behavioral and Brain Sciences, 31*, 357-378.

Park, K. M. (2001). In search of an Asian identity in mathematics education. *Math Love, 25*, 143-144.

Park, K. M. (2004). Factors contributing to Korean students' high achievement in mathematics. In Korea sub-commission of ICMI, *The report on mathematics education in Korea*, 85-92.

Polya, G. (1954). *Mathematics and plausible reasoning: Volume I.* Princeton, NJ: Princeton University Press.

Richland, L. E., Morrison, R. G., & Holyoak, K. J. (2006). Children's development of analogical reasoning: Insights from scene analogy problems. *Journal of Experimental Child Psychology*, *94*, 249-271.

Richland, L. E., Holyoak, K. J., & Stigler, J. W. (2004). Analogy use in eighth-grade mathematics classrooms. *Cognition and Instruction, 22*, 37-60.

Sfard, A. (2001). There is more to discourse than meets the ears: Looking at thinking as communicating to learn more about mathematical learning. *Educational Studies in Mathematics*, *46,* 13-57.

Sfard, A. (2008). *Thinking as communicating: Human development, the growth of discourses, and mathematizing.* Cambridge, UK: Cambridge University Press.

Sriraman, B. (2006). An Ode to Imre Lakatos: Bridging the Ideal and Actual mathematics classrooms. *Interchange: A Quarterly Review of Education, 37*(1&2),155-180.

Sternberg, R. J. (1977). *Intelligence, information processing, and analogical reasoning: The componential analysis of human abilities*. Hillsdale, NJ: Lawrence Erlbaum Associates.

Walshaw, M., & Anthony, G. (2008). The teacher's role in classroom discourse: A review of recent research into mathematics. *Review of Educational Research, 78*(3), 516-551.

Chapter 11

Students' Reasoning Errors in Writing Proof by Mathematical Induction

Alwyn Wai-Kit PANG Jaguthsing DINDYAL

Although mathematics is essentially deductive, students are exposed to various other types of reasoning when studying mathematics in school. They do not always understand the subtle differences in the types of reasoning that they come across and whether a particular statement based on a type of reasoning is valid or not. This invariably leads to reasoning errors. In this chapter, we explore some of the errors made by students in writing proofs by mathematical induction from a sample of 69 Junior College (Year 12) students. The data collected included students' responses to a test and post-test interview with selected students. The common errors were then further analysed in the light of the literature to understand the mathematical reasoning behind these errors. Semi-structured interviews were conducted with 11 selected students based on the solutions they had provided to either, confirm, refine or reject the initial understandings.

1 Introduction

Mathematics is a hierarchical subject in which apparently simple ideas are in fact very abstract. "Mathematics has developed into an extensive hierarchy or network of concepts, each more abstract than, and dependent upon, those feeding into it" (Barnard, 1996, p. 7). What cements the different abstract ideas in mathematics is reasoning and more specifically deductive reasoning, the basis of proofs in mathematics. There are various methods of proofs in mathematics but the method of proof that most students encounter at the post-secondary level (grades 11 and 12) is proof by mathematical induction or simply proof by induction. Proof by mathematical induction should not be confused with the common meaning of the term induction and inductive reasoning.

> There is an unfortunate ambiguity in the word 'induction'. On the one hand, the inductive method is a heuristic method for arriving at a conjectured generality describing a finite sequence of examples. On the other hand, mathematical induction is a rigorous form of deductive proof (Ernest, 1984, pp. 180-181).

1.1 *Proof by Mathematical Induction (PMI)*

PMI is a popular proof technique in discrete mathematics and number theory at higher levels. The method of the proof is based on the properties of natural numbers described as follows: if $P(n)$ is a proposition that is defined for every natural number n and $P(1)$ is true and also $P(k + 1)$ is true whenever $P(k)$ is true then $P(n)$ is true for all natural numbers n. Thus, a proof by PMI involves the following (see Ernest, 1984, p. 176): (1) proof of $P(1)$ as true, called the "basis", (2) assume $P(n)$ is true, called the "inductive hypothesis", and (3) proof of $P(n+1)$ as true from the inductive hypothesis mentioned in (2). Harel (2002) terms (1) above as the "Basic step" and (3) as the "Inductive step" (p. 189). Harel also proposed two categories for the problems in standard textbooks:

- Recursion problems which require the formation of a recursive representation of a function. For example: Prove that $1 + 3 + 5 ... + (2n - 1) = n^2$, can be expressed by the rule, $f(n) = f(n-1) + (2n-1), f(1) = 1$.
- Nonrecursion problems are those that do not involve the recursive representation of a function. For example: Prove that $n < 2^n$ for all positive $n > 3$. (p. 190)

In this chapter, we report students' reasoning errors when writing proof by mathematical induction. More specifically we look at the nature of the students' errors and why they made such errors in the proof by mathematical induction which involved a nonrecursion problem.

2 Methodology

This is a qualitative study that was carried out in a Junior College (JC) in Singapore. A qualitative approach was used, as qualitative studies seek to gain understanding from the observed data and build towards a theory from the intuitive understandings gained (Merriam, 1998). The participants in this study were 69 second year (Year 12) JC students spread over four different classes. These students had finished the JC curriculum and were preparing for their Preliminary Examinations, which is a school summative assessment at the end of the JC course and in which students are tested on what they are taught over the two years.

There were 28 female and 41 male students in the sample. These students were about 18 years of age and had all taken Additional Mathematics, a harder O-level course (at the end of Year 10) in their respective secondary schools. All of the students who participated had to sit for a test which lasted two hours. The written test which was constructed included 15 free response items, covering a range of topics which included proof by mathematical induction, inequalities, transformation of graphs, functions, sequences, calculus, trigonometry, complex numbers, permutations and combinations. These topics were chosen to represent a spread of the A-Level Pure Mathematics syllabus. Questions were constructed to test mathematical reasoning skills, not their ability to manipulate expressions. The test items were carefully selected and checked by an expert in the field. The instrument was piloted with a small group of 6 students (other than those in the sample) and as a result some of the questions were revised. The test was marked and the errors documented. Here, we only report the errors that the students made on Question 2 of the test in which they had to use PMI. The common errors were then further analysed in the light of the literature to understand the mathematical reasoning behind these errors. Semi-structured interviews were conducted with 11 selected students based on the solutions they had provided to either confirm, refine or to reject the initial understandings. This required an insight into the students' thought processes and an understanding of the meaning that the students had constructed for themselves. Six of the interviewed students were from the top third, two from the middle and three from the bottom third of the students in the sample.

The Principle of Mathematical Induction which was taught to the students from the Junior College where this study was conducted had the following structure:

(i) Defining the $P(n)$ statement.
(ii) Verifying the Base Step of the induction (the verification that statement $P(1)$ is true).
(iii) Proving the Induction Step that if $P(k)$ statement is true, then the $P(k+1)$ statement is true.
(iv) Concluding the proof of mathematical induction with a statement.

The problem on mathematical induction that was used in this instrument was a nonrecursive type of problem (see Harel, 2002). The students were required to prove the nth derivative of a function (See Figure 1).

Question 2

Prove by induction that for all positive integers, $\frac{d^n}{dx^n}(xe^x) = (x+n)e^x$.

Figure 1. Question 2 of test instrument

3 Results and Discussion

As observed by Ernest (1984), a successful proof by mathematical induction hinges on the basic step and the induction step. These skills are in turn dependent on various sub-skills. The ability to prove the base case of the induction presupposes the ability to verify that fixed numerical properties hold for particular case $P(1)$. This in turns depend on one's ability to perform substitution into algebraic expression in a single variable. The ability to prove the induction step depends on the ability to prove an implication statement (i.e. a "if... then..." statement) by deducing a conclusion from the hypothesis. It consists of the ability to make deductions from algebraic identities which in turn depend on one's skill in manipulating algebraic expressions and identities. Baker (1996) reported that many students faced difficulties with mathematical induction because of a lack in mathematical content knowledge. They were unable to operate with symbols and lacked the skills which they were supposed to have mastered previously.

3.1 *What are the mathematical reasoning errors made by students?*

Students in this study seemed proficient in algebraic manipulation: only four of students in this study were hampered by algebraic errors. No student made mistake in the verification of the base case. However, 36 of the participants wrote statements which suggested that they did not truly understand the reasoning behind the proof by mathematical induction. For example, eight of them had stated in their conclusion that the proof by mathematical induction established the truth of a statement $P(n)$ for all *real* numbers, n.

We shall now discuss some of these errors. First, many participants understood mathematical induction to be a procedure. When asked what mathematical induction meant to them, these participants narrated the steps they have to perform for the mathematical induction as exemplified in the following interview excerpt of JT (I = Interviewer):

I: Can you explain what mathematical induction is to you?

JT: How by assuming the Left Hand Side [LHS] and Right Hand Side [RHS] is equal each other than… eh… that the assumption will be true, or that particular equation. So let say the beginning... eh. For me, I do it because this is the formula, it is always this structure, so I just plainly follow this structure and derive my answer.

I: So what is the proof by induction, or rather what are you trying to establish? Are you trying to establish anything?

JT: Establish that by… erm... if $P(1)$ is true, if $P(k)$ is true then all the rest should be true.

I: Can you explain your working to me?

JT: First I used this as my assumption.
I: The $P(n)$ statement, you mean?
JT: Ya. After using $P(n)$, by plugging n as 1, try to prove that LHS = RHS. Once this is found out, then I can say that $P(1)$ is true because it satisfy the equation on both sides. So after this we go on to k. By changing n to k, k can be any random … er… is a variable, so this will be the hypothesis. If we can prove that $P(k + 1)$ is true then $P(k)$ should be true.

Similar answers depicting the procedural nature of students' thinking have been collected by Baker (1996). In his study on the difficulties faced by students in mathematical induction, Baker also found that students tend to focus their attention on procedures rather than on concepts or applications. This should not be surprising given the dominance of instrumental learning among students. In the following interview with TP (a student in the top third of the group), we see not only a procedural description of mathematical induction, but also a confession of the absence of relational understanding.

I: Can you explain to me how proof by mathematical induction works?
TP: First you must let $P(n)$ be… write down the whole question. Then you prove $P(1)$ true, then you assume $P(k)$ true for something. Then you consider the $P(k + 1)$, then must try and get… say if I put $(k + 1)$ into the right hand side. Then if it is true then all true.
I: Why does it work? It is a proof. Why does this process work?
TP: I don't know.
I: You don't know?
TP: They tell me to do like that, so I do like that. This is the format for induction.

A consequence of instrumental learning habits is the formation of external conviction proof schemes: namely, the authoritarian, symbolic and ritual proof schemes (Harel & Sowder, 1998). Here, we see evidence of the authoritarian proof scheme. TP's knowledge was based on "they tell me to do like that, so I do like that". Besides, TP had also justified his argument by appealing to the "format of induction" — an indication of the ritual proof scheme. The ritual proof scheme judges the correctness of the proof based on the form of the argument (Sowder & Harel, 1998). The second half of our interview with TP further illustrated this. We showed TP the following working (Figure 2).

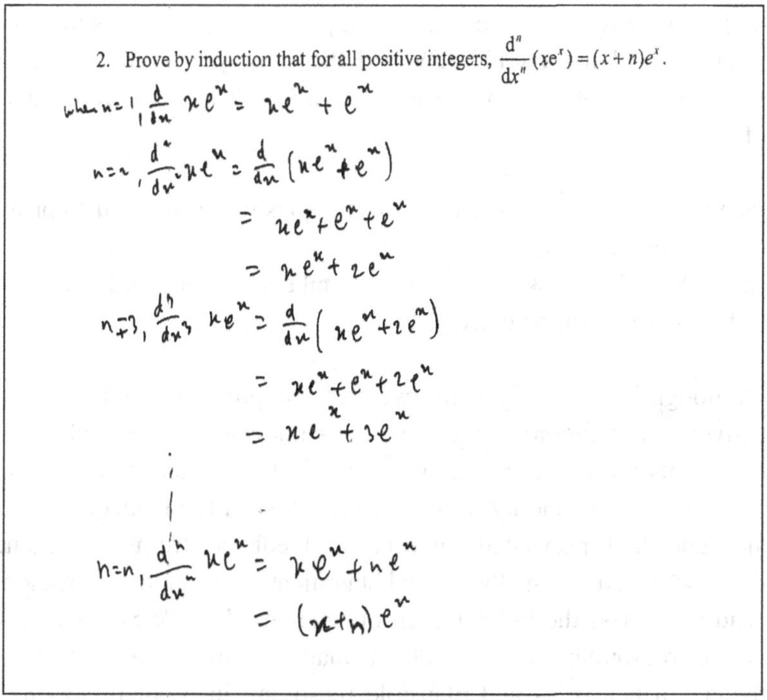

2. Prove by induction that for all positive integers, $\frac{d^n}{dx^n}(xe^x) = (x+n)e^x$.

when $n=1$, $\frac{d}{1dx} xe^x = xe^x + e^x$

$n=2$, $\frac{d^2}{dx^2} xe^x = \frac{d}{dx}(xe^x + e^x)$
$= xe^x + e^x + e^x$
$= xe^x + 2e^x$

$n=3$, $\frac{d^3}{dx^3} xe^x = \frac{d}{dx}(xe^x + 2e^x)$
$= xe^x + e^x + 2e^x$
$= xe^x + 3e^x$

$\vdots$

$n=n$, $\frac{d^n}{dx^n} xe^x = xe^x + ne^x$
$= (x+n)e^x$

Figure 2. Inductive proof scheme

I: In that case, why don't you do this?
TP: Because this is not induction.
I: Why is this not induction?
TP: Induction is that one (pointing to the correct answer).
I: Why is induction that one and not this one?
TP: Because we learn this one.
I: If you are the teacher will you mark it correct or wrong?
TP: Wrong.

I: Why?
TP: I mean it is correct, but it is not induction.

TP had presented his proof by mathematical induction correctly, but when asked why he could not have presented it inductively, TP said that the inductively presented answer was wrong because that is not the "format of induction". TP had thought that an argument using inductive reasoning alone was also acceptable as a proof except that it is not "proof by mathematical induction". KW had the same perception as TP. We showed KW the solution in Figure 2 and asked if it was acceptable as a proof.

KW: It is not acceptable because the question wants you to prove it by induction.
I: So if I just ask you to prove — take away the word induction?
KW: Then it will be correct.

Although in this study, only two students presented their arguments inductively as in Figure 2, the interviews uncovered others who would accept an inductive argument as a proof. This is an example where instrumental understanding of a topic had masked the inadequacy of the learner and had prevented appropriate feedback from the teachers. Students who validate mathematical statements with inductive reasoning are said to possess the inductive proof scheme (Harel & Sowder, 1998). Inductive reasoning is essential in mathematics. It is a necessary ingredient in heuristics and plausible reasoning in problem solving. In practice, mathematicians do not use only purely logical deductions but an interplay of both inductive and deductive reasoning (Polya, 1957). The concern here is not that students employ inductive reasoning, but that students do not develop beyond the inductive proof scheme, and do not appreciate that a mathematical proof must be rigorous, general, complete and conclusive. A student who possesses the inductive proof scheme has a limited understanding of mathematical proofs.

It could be due to their limited exposure to proofs that students have difficulty coping with proofs. Like TP and KW, they do not know that in demonstrative reasoning (as in the formal communication of mathematical proofs) all arguments must be deduced from definitions or known results by logical inference. Constructing a proof requires students to appreciate and adhere to mathematical logic, especially implication (i.e., *modus ponens*). Implication is central to mathematics since it is the logic construct that holds mathematical statements and ideas together. It is the key to understanding the induction step. However as we see in the following interview, students are often confused between "if… then…" and "if and only if" statements. They may in fact think that the two statements are one and the same. Such misconception was termed "child's logic" by O'Brien, Shapiro and Reali (1971).

I: What is the meaning of your conclusion?

CS: I think, since I have gotten $P(k + 1)$ is true that means my assumption actually is true, then therefore $P(n)$ is true.

I: Why is it that if you prove $P(k + 1)$ is true, then your assumption is true?

CS: Because…

I: But you assumed $P(k)$, you prove that $P(k + 1)$ is true, what have you proven? You have only proven that $P(k + 1)$ is true. Does it then show that $P(k)$ is true?

CS: But I prove $P(k + 1)$ based on $P(k)$, so if $P(k + 1)$ is true that means $P(k)$ should also be true, right?

I: You assumed it true initially?

CS: If I can't prove $P(k + 1)$ is true that means $P(k)$ is not true. …… Since if I can't prove $P(k + 1)$, that means actually my $P(k)$ got something wrong. Since I have proven $P(k + 1)$ is true wouldn't $P(k)$ be true also?

CS thought that the proof of the $P(k + 1)$ statement validates the assumption that $P(k)$ is true. Students do not regard the induction step to mean "if $P(k)$ is true then $P(k + 1)$ is true", but rather "to prove that $P(k + 1)$ is true". Students do not understand that the word "assume" in mathematical logic actually meant "given that the statement is true". JT felt that one had to "assume because it is not yet true". When pressed harder what that meant in other places, JT said it was based on memory.

I: You wrote here assuming $P(k + 1)$ is true. What do you mean by that statement?

JT: Because it is not true yet, so we must assume.

I: You also wrote assuming $P(1)$, what is the rationale for writing these two statements?

JT: Before that, I remembered that there is the word assume, so I don't know where to put, so I wrote it there. And it seems logical to me.

Upon further investigation, JT revealed the belief that one can only assume that something is true when there is a strong case that it is going to be true. So the base case in the proof by mathematical induction functions as the basis for assuming statement $P(k)$ to be true.

I: If I don't prove $P(1)$ is true, would the proof by mathematical induction work?

JT: I think it wouldn't.

I: Why wouldn't it work?

JT: Because if it doesn't work for 1 then it should not work for k... I think this first step is quite essential.

I: Why is that so?

JT: Because before we create an assumption of a variable, by putting 1 as the first integer in, I think it is the basis where once $P(1)$ can be proven then we can go on to assume other equations.

The base case is thought to be, in JT's words, "the basis where once $P(1)$ can be proven then we can go on to assume other equations". JT is not the only one who used the word "basis" to refer to $P(1)$. AGC (in the top third of the group) thought that without putting $P(1)$ in front, the proof of the induction step "if $P(k)$ is true, then $P(k + 1)$ is true" is "illogical" because then we are "assuming something and you try to prove it later".

I: Can you assume $P(k)$ first, prove $P(k + 1)$, then prove $P(1)$ true later? Is it possible?

AGC: Hmm... I mean on paper it is correct, but it is not logical... because when you assume $P(k)$ is true, you do not have any basis that $P(k)$ is true. It is like assuming something and you try to prove it later when you have done everything already... it is illogical.

KW used the word "basic benchmark" to describe $P(1)$ without which one "cannot assume $P(k)$ is true".

I: Must you always prove $P(1)$ first?

KW: If you don't prove $P(1)$ first then you won't have a basic benchmark. Then you cannot assume $P(k)$ is true.

Linking the base case to the induction step is a key step to the understanding of the proof of mathematical induction. Baker (1996) had noted that students had trouble relating the base case of the proof to the induction step. Some of the participants in this study, like JT, AGC, KW, had the misconception that the base case must always be done before the induction step. CBS had correctly answered this question on mathematical induction, but felt that while it may be acceptable to have the base case placed at the end of the proof, it is not conceptually correct to do so; in CBS words, "the concept is not there", and he provided the reason for it.

I: Can you explain to me why the procedure of mathematical induction actually proves a statement.

CBS: Because... induction is like trying to prove a statement by like ... trying the data. It is like domino effect. So you want to test the first data, as if like the first domino... you want to use, and you want to do the subsequent domino by applying a general formula. Then... in that case... one general formula affecting the next... subsequent... data can be proven. That is what induction is all about.

I: Why is $P(1)$ necessary? Why must we prove $P(1)$?

CBS: That is the starting point. If you didn't start the first, you don't know if there is any limit to the bottom.

I: If you did not prove $P(1)$ first, say you assume $P(k)$ prove $P(k + 1)$ first then later you go back to prove $P(1)$. Is that possible?

CBS: It is possible but in terms of...what do I call it... the way you think it is not working. As in... because you are pushing the domino from the beginning to the end and not from the end to the beginning. Something like that.

I: Somebody actually wrote this.

We showed the working in Figure 3 to CBS.

I: Is this correct or wrong?

CBS: By right... must at least minus mark because the concept is not there. Conceptually it is like you are trying to push the 10th domino and because of that the first domino drop. Something like that. If that is the requirement of mathematical induction, if we just use it as a formula based proving method then you can accept, but conceptually it is not right.

I: So you think that $P(1)$ should go before $P(k)$?

CBS: Yes.

CBS seemed to have brought the analogy of the dominoes used to introduce the topic a little too far. The proof of mathematical induction presupposes an understanding of the defined properties of natural numbers and the recurrence and the ordering of the natural numbers (Ernest, 1984). Teachers usually use the analogy of the dominoes as mentioned by CBS or that of a man climbing up a ladder to illustrate it. However, in this study, eight of the students wrote that the induction establishes truth for all real numbers. As we found from the interviews there are other students who are at fault in this regard.

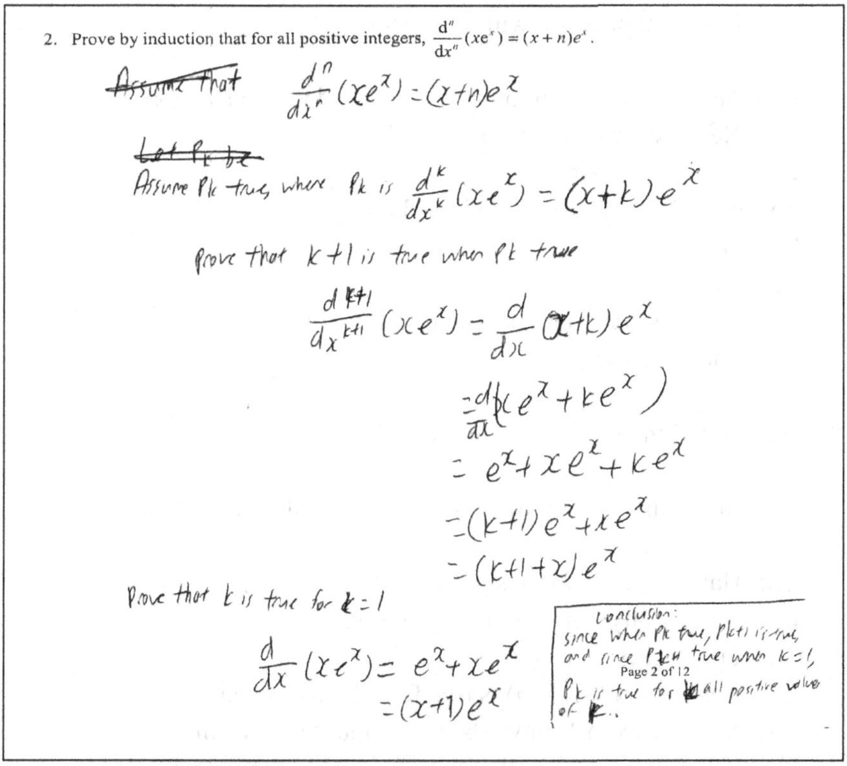

2. Prove by induction that for all positive integers, $\frac{d^n}{dx^n}(xe^x) = (x+n)e^x$.

~~Assume that~~ $\frac{d^n}{dx^n}(xe^x) = (x+n)e^x$

~~Let P_k be~~

Assume P_k true, where P_k is $\frac{d^k}{dx^k}(xe^x) = (x+k)e^x$

Prove that $k+1$ is true when P_k true

$$\frac{d^{k+1}}{dx^{k+1}}(xe^x) = \frac{d}{dx}(x+k)e^x$$
$$= \frac{d}{dx}(xe^x + ke^x)$$
$$= e^x + xe^x + ke^x$$
$$= (k+1)e^x + xe^x$$
$$= (k+1+x)e^x$$

Prove that k is true for $k=1$

$$\frac{d}{dx}(xe^x) = e^x + xe^x$$
$$= (x+1)e^x$$

Conclusion: since when P_k true, P_{k+1} is true, and since P_{k+1} true when $k=1$, P_k is true for all positive value of k.

Page 2 of 12

Figure 3. An alternative presentation to mathematical induction

I: Is it possible that we prove by mathematical induction a statement for which it is true for all $n \in \Re$?

KK: Ya.

I: It is possible?

KK: Ya.

We showed AS the piece of work with a conclusion as shown in Figure 4.

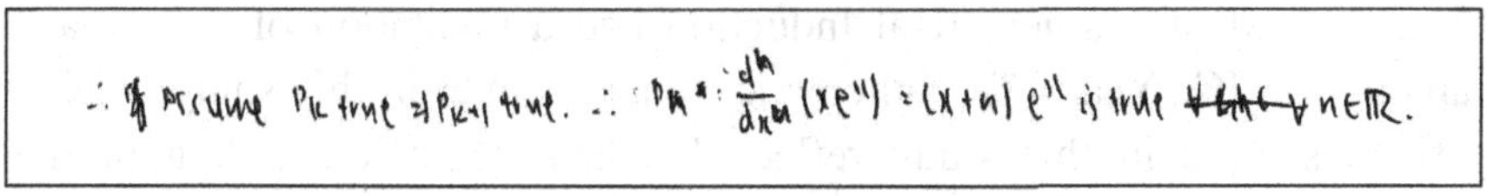

$\therefore$ If assume P_k true $\Rightarrow P_{k+1}$ true. $\therefore P_n: \frac{d^n}{dx^n}(xe^x) = (x+n)e^x$ is true ~~that~~ $\forall n \in \mathbb{R}$.

Figure 4. A wrong conclusion in mathematical induction

I: Is there a problem with this proof?
AS: No.
I: There is nothing wrong with it? Read the conclusion…
AS: Never say by mathematical induction.
I: Other than that?
AS: err…They never say $P(1)$ and all that…
I: What else?
AS: err…
I: Can this statement be true for all real values n?
AS: I don't know.

We showed CS the same piece of work (Figure 4).

I: Read the concluding statement. Is this conclusion right or wrong?
CS: Have to put in $P(1)$ is true.
I: Is there anything else that is wrong?
CS: I think it is correct.
I: It is correct? … $P(n)$ is true for all real. Is that correct?
CS: Must be from 1 onwards. $P(n)$ should start from 1.
I: If I change to "for all real ≥ 1", would it be correct?
CS: Ya. I think it is true.

As in Figure 4, such errors occur only in the students' conclusion and seemed minor, but it clearly revealed students' lack of understanding of mathematical induction. AS' and CS' replies also indicated a strong slant to the ritual proof scheme. They thought that the mistake was "never say by mathematical induction" or "have to put in $P(1)$ is true".

3.2 *What may be some causes of the errors?*

The method of Mathematical Induction had a reputation of being easy among the JC (Year 12) mathematics students. Yet in this study, 52.2% of the sample in this study reflected a lack of understanding of the method. So on the surface, the proof by mathematical induction may seem easy to master, but as the interviews show, it is a difficult concept to grasp. This may be because the method of proof presupposes other concepts which teachers assume students to know.

One such presupposition is the natural ordering of the natural numbers. We saw that in this study, more than 11.6% of the students had the misconception that the proof of mathematical induction validates the mathematical statement $P(n)$ for every real number, n. Students do not understand that the n in mathematical induction needs to be countable at the very least. They also need to know that besides being countable, the set of numbers generalised by n can be ordered. Students need a structural understanding of the set of positive integers.

Ernest (1984) suggested the use of analogies like the ascent of a ladder or the knocking down of dominoes to illustrate the relationship between the proof of mathematical induction and the ordering of the natural numbers. As we have noted in the interview with CBS, such analogies may also lead to confusion. That interview also showed that in using analogies, we need to carefully highlight the intent of the analogies. Students do deal with meanings, and when our instructional programs fail to develop the appropriate meanings in them, students create their own meanings and sometimes, meanings that are not appropriate at all (Davis, 1988). An example of this is the meaning given by students to the word "assume". In the case of JT, the word "assume" was written because the $P(k + 1)$ statement was not yet true. JT, KW, AGC and CBS all gave meaning to why the proof of the base step must be presented at the beginning – because it was needed for the assumption of the $P(k)$ statement. AGC particularly felt that if the base step is not established then the method of mathematical induction is "illogical" because then we are "assuming something and you try to prove it later". This misconception that mathematical induction is a method in which you assume what you have to prove and then prove it is one of the most common misconceptions among students (Ernest, 1984). Like CBS, students give their own meaning to keywords, format and solution structures if they are left unexplained.

The method of mathematical induction also presupposes that a student is familiar with logical implication and the notion of elementary proof (Ernest, 1984). The misconceptions which surfaced in the interviews with JT, KW, AGC and CBS demonstrated that students may not understand implication, or rather the proof of an implication statement through the assumption of its antecedent (Ernest, 1984). These students regard the induction step solely as proving the $P(k + 1)$ statement and not proving that "if $P(k)$ is true then $P(k + 1)$ is true". O' Brien, Shapiro and Reali (1971) advised that in dealing with mathematical issues involving conditional statements, teachers should not take for granted students understanding of implications expressed in the "if… then…" language. Due to the lack of exposure to proofs, and as we have seen from the errors, students at Junior College level still do not have an understanding of logical implication, much less, an understanding of the meaning of the "$\Rightarrow$" symbol.

Baker (1996) termed inductive reasoning everyday reasoning. One possible reason for the pervasiveness of the inductive proof scheme could be the common use of inductive reasoning in everyday life. The occurrence of events in everyday life is probabilistic in nature so past observations are often used to make generalisations on the nature of the occurrences. But mathematics is a deductive science and cannot accept inductive reasoning as a valid form of proof. Moreover in everyday reasoning, if there are reasonable-enough justifications from past experiences, a conjecture is often accepted as truth. This is contrary to providing a mathematical proof where a conjecture is not considered to have the same rigour of truth as a proven theorem. Ernest (1984) suggested that the prevalence of the inductive proof scheme could be due to the confusion over the name induction. Ernest also suggested that it is important for teachers to carefully highlight the distinction between heuristic inductive reasoning and the proof by mathematical induction as a rigorous deductive proof. He further suggested that to distinguish the two, the heuristic inductive reasoning be called by some other name like method of generalisation other than induction. This may not be a problem in Singapore, because students are not told that the heuristic inductive reasoning method is also known as induction. So when Junior College students mention the word "induction" it is commonly taken by the teachers and students to refer to mathematical induction. It is more likely that the exhibition of the inductive proof scheme by students here is due to their misconception that the generalisation from patterns constitutes a valid justification. This, as mentioned, may be a carry-over effect from everyday reasoning, or could be due to their prior experience with questions that require students to deduce an answer from patterns, which they have encountered in secondary schools.

JT is one student who had accepted the inductive proof scheme as a valid proof because of his prior experiences in secondary school. I showed JT Figure 2 and asked,

I: Is this justifiable? Does this prove?
JT: Hmm... It is logical. I think if I can think of it, I will also do this.
I: You also will do this?
JT: Why not? Because in secondary school, it is also like that. There is always... this kind of question is always the last part. If you haven't learned how to do Mathematical Induction [MI], then... this will be... ya... its very possible.
I: You think you can prove something by the recognition of patterns?
JT: Ya... patterns... because Maths is about patterns...

JT does not appreciate the difference between formulating a conjecture and a proof "because in secondary school, it is also like that" and "because Maths is about patterns". Harel and Sowder (1998) noted that the application of inductive reasoning in mathematical problem solving is essential. In elementary mathematics, students are encouraged to explore, often without the need for rigour. An example would be the pattern explorations they do in secondary schools. At that level, once students are able to come out with the general mathematical statement from the observed patterns, the problem is considered solved. But some students have come to accept the conjecture as the final product. They do not realize that this is not so in a proof. The demand in a formal proof is much greater. As students advance from merely stating their conjecture to producing formal mathematical proofs, they need to shift from generalising from the purely inductive heuristic methods to deriving new mathematical objects from definitions through a sequence of logical deductions. They need to know that on top of merely exploring mathematics, they are constructing mathematics based on valid justifications. Students' mathematical justifications must develop beyond the inductive proof scheme.

In Singapore, the proofs which students do (up to secondary school) consist mainly of simple validation of identities where they start from the left-hand side [LHS] and arrive at the right-hand side [RHS]. They are not familiar with the language of a mathematical proof. Mathematical proof is a new experience. In fact, from our experience, many students think that a proving question is similar to any other question — the only difference being that the answer is given when they are asked to prove. As Baker (1996) observed, one missing component from previous studies on proof by mathematical induction was that prior experiences of students have only been given limited attention. The inability to reconcile the rigour required in a proof (in this case mathematical induction) with their prior experiences of explorations (of patterns) may be another reason for the prevalence of the ritual proof scheme. As TP said, the proof in Figure 2 "... is correct, but it is not induction". Unable to resolve the difference, TP accepted the proof of mathematical induction as a method — "they tell me to do like that, so I do like that. This is the format for induction." Mathematical induction is a new experience; it is not a generalisation of previous more elementary experience (Ernest, 1984).

Finally, Harel and Sowder (1998) noted that the kinds of problems typically introduced to students in their first experience with mathematical induction are cognitively inadequate. Ernest (1984) noted that the frequent use of examples involving finite series in mathematical induction has led to a common misconception concerning the form of the $P(k + 1)$ statement in the proof. In such cases, students simply assume that the $P(k + 1)$ statement equals the expression given in the $P(\mathrm{k})$ statement "plus" the $(k + 1)^{\text{th}}$ term. Students derive this habit from mindlessly doing many mathematical induction questions where they start to prove the $P(k + 1)$ statement with $\sum_{r=1}^{k+1} u_r = \sum_{r=1}^{k} u_r + u_{k+1}$. It is in the light of such observed errors that for study we have specifically chosen a proof item which involves differentiation and not summation. Not surprising, similar errors also surfaced in this study. Figure 5 shows two solutions from participants who wrote in their $p(k+1)$ statement, "$p(k)$ + another term".

Testing P_{k+1} ie to prove true " $\frac{d^{k+1}}{dx^{k+1}}(xe^x) = (x+k+1)e^x$ "

LHS $= \frac{d^{k+1}}{d^{k+1}}(xe^x)$

$= (x+k)e^x + \frac{d}{dx}(xe^x)$

$= (x+k)e^x + e^x(x+1)$

Assuming ~~taken~~ P_{k+1} is true, on $\frac{d^{k+1}}{dx^{k+1}}(xe^x) = (x+k+1)e^x$

LHS: $\frac{d^{k+1}}{dx^{k+1}}(xe^x) = (xe^x + e^x) + (x+k)e^x$

$= (x+k+1)$

Figure 5. Analogical reasoning errors in Question 2

AS produced the first of the above solution and during the interview and we asked:

I: Why did you write this, the line after LHS?

AS: It is… You know last time, for other things, we say the k term plus the $(k + 1)$th term. So this is the k term and for the $(k + 1)$th term, I just differentiate this…oh, how come I differentiate this… I am kind of lost now.

I: Is it always true for mathematical induction that to prove $P(k + 1)$, is it always true that you must have the kth term, or rather have what you have assumed earlier plus the $(k + 1)$th term?

AS: Errr… Ya.

This is an analogical reasoning error where students simply use a procedure based on the superficial features from one context on a different problem.

Harel and Sowder (1998) suggested that students' difficulties in mathematical induction may be attributed to teachers' hasty introduction of the formal expression of this principle to the students. This observation is very true. To help students with the topic, teachers provide them with a specific structure to follow. This may be deemed "successful" from the high percentage of "correct" answers observed. However as Ernest (1984) acknowledged, such a method of instruction lend itself to an instrumental understanding of the proof. Although teachers do try to have students understand the rationale behind the steps, the structure emphasized the method rather than the logical reasoning and encouraged the development of the ritual proof scheme. The reason why the ritual proof scheme may be common among students may be because teachers over-emphasized the presentation of solutions, and students try too hard to adhere closely to a stipulated format. Another drawback to hastily giving students a formal expression is that students may focus wrongly on the unimportant points of the presentation.

Although this was not part of the current study, one issue with student difficulty in writing and understanding proof may be the teachers own difficulties with the notion of proof. Stylianides, Stylianides and Philippou (2007) highlighted the ideas of teachers' *fragile* knowledge and *solid* knowledge of proof: "A prerequisite in the effort to make proof central to school mathematics is that teachers of all levels have *solid* knowledge of proof, that is, sturdy knowledge that withstands to inject contradictions into it" (p. 146). These researchers working with preservice teachers found some errors very similar to those that were found in this study.

4 Conclusion

Our aim in this chapter was to identify the nature of students' errors and the reason behind these errors when writing proof by mathematical induction. The problem chosen in this study was of the nonrecursion type involving the nth derivative of an exponential function. Reasoning errors were found in both the basic step and the induction step. In addition, some students who otherwise gave mathematically correct answers showed a lack of understanding of proof by mathematical induction when we showed them the erroneous solutions of their friends. First, although the basic step which involved the proof of $P(1)$ was done correctly, students followed a ritual proof scheme that this part had to precede the inductive step. For example, CBS did not think that the proof in Figure 3 was correct. Second, regarding the inductive step, some students after having proved that $P(k+1)$ is true by assuming that $P(k)$ is true, believed that this implied that $P(k)$ was true. Other errors included stating that the $P(n)$ statement is true for all real values of n. This could be observed in the interview with AS where the mistake in the conclusion was thought to be "never say by mathematical induction" or "never say $P(1)$ and all that" and not because the proof by mathematical induction cannot be used to establish the truth of a statement $P(n)$ for all real numbers, n.

PMI is certainly logically and conceptually challenging even for undergraduate students and hence does it make sense to teach it to younger students? There are two options: we can argue for its removal from the Year 12 level content or we can maintain its current status with an emphasis on simple recursion and possibly some nonrecursion types of problems. While the *strong induction* version of PMI can be even more challenging for Year 12 students, the former option may be too drastic in that it may be the only opportunity for some students to learn about one fundamental aspect of proof in mathematics. We shall argue for the second option with emphasis on selected types of recursion and even nonrecursion problems. A few points for teachers to consider when teaching about PMI:

- Avoid an over emphasis on presenting a sequence of steps in the writing of the proof.
- Analogies such as "dominos knocking one another" or "climbing up the ladder" have to be carefully assessed (see Ernest, 1984).
- Many errors occur because of a poor understanding of the base step and the inductive step. The validity of the proof of $P(n)$ for natural values of n as opposed to real values of n should be carefully discussed with students.
- Recursive types of problems should be used prior to using the nonrecursive type (see Harel, 2002) as they tend to be easier for students. However, a sole reliance on the former type of

problems gives a limited view of PMI, particularly in the use of the inductive step.

- Correct solutions from students should be explored further to uncover deep-rooted misconceptions. As this study demonstrates, some students with correct PMI solutions were not very sure when they were shown erroneous solutions.

To conclude, students' errors in PMI provide teachers with an excellent window into the students' thinking and understanding of the proof. An analysis of such errors can help teachers to strengthen their pedagogical content knowledge for teaching this method of proof.

References

Baker, J. D. (1996). *Students' difficulties with proof by mathematical reasoning*. Paper presented at the Annual Meeting of the American Educational Research Association, New York.

Barnard, T. (1996). Teaching proof. *Mathematics Teaching, 155*, 6-10.

Davis, R. (1988). The interplay of algebra, geometry, and logic. *Journal of Mathematical Behavior, 7*, 9-28.

Ernest, P. (1984). Mathematical induction: A pedagogical discussion. *Educational Studies in Mathematics, 15*(2), 173-189.

Harel, G. (2002). The development of mathematical induction as a proof scheme: A model for DNR-based instruction. In S. R. Campbell, & R. Zaskis (Eds.), *Learning and teaching number theory: Research in cognition and instruction* (pp. 185-212). New Jersey: Ablex Publishing Corporation.

Harel, G., & Sowder, L. (1998). Students' proof schemes: Results from exploratory studies. In A. H. Schoenfeld, J. Kaput & E. Dubinsky (Eds.), *Research in collegiate mathematics education III* (pp. 234-283). Providence, RI: American Mathematical Society.

Merriam, S. B. (1998). *Qualitative research and case study applications in education*. San Francisco, CA: Jossey-Bass.

O' Brien, T. C., Shapiro, B. J., & Reali, N. C. (1971). Logical thinking — Language and context. *Educational Studies in Mathematics*, *4*(2), 201-219.

Polya, G. (1957). *How to solve it. A new aspect of mathematical method.* (2nd Edition). Princeton, NJ: Princeton University Press.

Sowder, L., & Harel, G. (1998). Types of students' justifications. *Mathematics Teacher*, *91*(8), 670-675.

Stylianides, G. J., Stylianides, A. J., & Philippou, G. N. (2007). Preservice teachers' knowledge of proof by mathematical induction. *Journal of Mathematics Teacher Education*, *10*, 145-166.

Chapter 12

Presenting Mathematics as Connected in the Secondary Classroom

LEONG Yew Hoong

The pressure to complete syllabus within a limited time in the work schedule of a teacher can sometimes lead one to view mathematics "content" as comprising points from a check list to tick against. This can in turn lead to teaching as "covering" isolated pieces of skills practice. Against this portrait of instructional practice is one where mathematics is presented as skills and concepts that are closely connected to each other. In this conception of classroom practice, students are encouraged to see the mathematics that they are learning as situated and connected to a wider context of mathematical knowledge. In this chapter, we discuss ways in which this alternative portrait of mathematics instruction can be enacted, drawing upon examples that include findings from a project that places a high value on presenting mathematics as being connected.

1 Introduction

The important role that connection-making plays in the teaching and learning of mathematics is perhaps best summarized by Healy and Holyes (1999): "We start from the position that mathematical meanings are developed by forging connections between different ways of experiencing and expressing the same mathematical ideas" (p. 60). Over the last decade, a number of studies were carried out to explore the different aspects of connection-making in mathematics instruction (e.g., Bosse, 2003; Horton, 2000; Knuth, 2000).

In Singapore, there have also been recent developments that are in keeping with the emphasis on the importance of making connections in mathematics teaching and learning. This is more apparent in the pentagonal model (see Figure 1) which is the framework of the Singapore Mathematics curriculum for Years 1 to 12. On the side of the pentagon labeled Processes, "connections" is inserted additionally as one mathematical process that is valued in the latest mathematics curriculum revision in 2006.

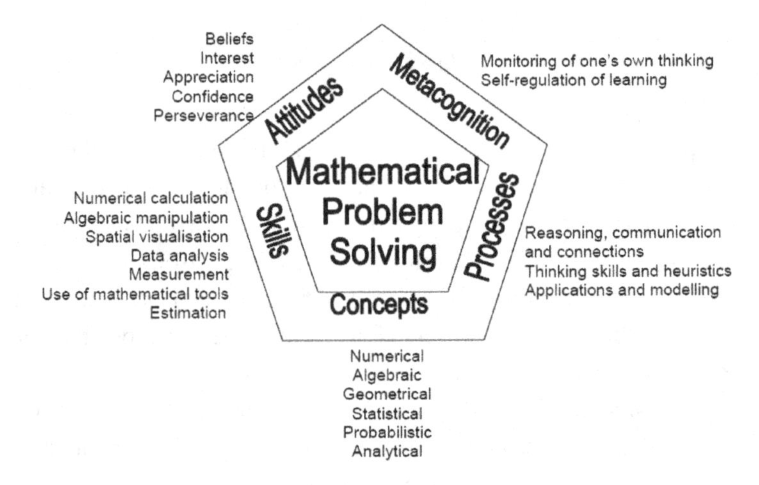

Figure 1. Singapore Mathematics curriculum framework (MOE, 2006, p. 2)

According to the Mathematics syllabus of the Ministry of Education (MOE) in Singapore, "[c]onnections refer to the ability to see and make linkages among mathematical ideas, between mathematics and other subjects, and between mathematics and daily life. This helps students make sense of what they learn in mathematics" (MOE, 2006, p. 4). This is a broad vision of "connections", and includes not only making links within the discipline of mathematics, but also connecting to domains of knowledge outside the traditional sphere of mathematics. For the purpose of this chapter, I restrict my discussion on the former category of intra-mathematics connections.

This vision of teaching mathematics as connected is perhaps upheld against a contrastive portrait of teaching that is deemed as more prevalent in Singapore classrooms. In this latter image of teaching, mathematical facts and procedures are broken down into bits, taught in isolation, and learnt separately. Such a way of teaching mathematics encourages the learning of small bite-size ideas but downplays the motivation to see the big overarching ideas that tie the bits together; ultimately, this form of instruction, when sustained, leads to a distorted idea in students of the nature of mathematics: Instead of a tapestry of an interconnected and consistent system of knowledge, it is a memorisation and practice of a set of unrelated rules and formulas.

The portrait of classroom teaching that emphasises connections, on the other hand, presents a different picture. It stresses not only the learning of mathematical skills and facts but also the organising conceptual frames that bind these ideas together. In lessons of this nature, the focus is not only on zoomed-in details of procedures and local reasoning but also in the zoomed-out views of how and why these details work in relation to neighbouring and overarching mathematical conceptual fields. For this paper, I focus on some features of classrooms that subscribe to making mathematical connections an integral part of learning: Teachers make explicit connections between modes of representation to help students see the intermodal links more easily (Visual connections); also, mathematical ideas are consciously weaved together (Idea connections); and, such efforts to make Visual and Idea connections are not carried out only as one-off occurrences, rather, they are sustained across time over lessons (Temporal connections).

2 Visual Connections

Teachers do not just engage students verbally; they also use representations — in various media forms such as drawings on the whiteboard or projections of computer screens — to help students focus visually on the discussion track and sometimes even to provide hints to attack routes to solving mathematical problems. There are many visual representational forms commonly used in the mathematical domain: number lines, tables, Venn diagrams, matrices, graphs, among others. One aspect to the work of teaching is to harness these representations appropriately to advance the mathematical agenda in the classroom.

Sometimes, within a unit of instruction, more than one mode of representation may be presented to highlight the transition between different forms of representation (for example, from the concrete representation, to the diagrammatic form, and then the algebraic mode). In this case, quality teaching is not merely seen by the mere use of these representational modes, but also in the careful linking of these modes in practice so that the intended smooth intermodal transitions are effected.

I present below an example of how these modes can be harnessed in a tightly-linked fashion on the whiteboard in the context of this problem that is common in Year 7 Singapore textbooks: Peter is 7 years older than John. Given that the sum of their ages is 33, find their respective ages. The teacher can start by deconstructing the conditions in the problem under the "textual mode" of representation shown in the first panel of the whiteboard; this can be followed by the corresponding model method visuals under the "diagrammatic mode" in the second panel; finally, the algebraic solution can be presented under the "Symbolic mode" in the last column of the board. Figure 2 illustrates the various modes and their intermodal links in the context of the problem.

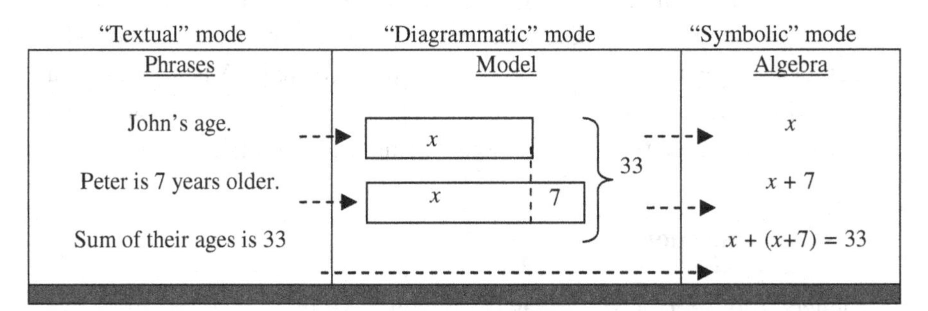

Figure 2. Presentation of links among modes, extracted from Leong (2006)

This brief discussion on Visual connections is to highlight its importance in mathematics learning. For a longer discussion on external visual representations, the reader may refer to Lowrie's (2011) chapter in this volume. Often, teachers are aware that students are comfortable in a particular mode of representation. In the case of the example presented, Year 7 students tend to be more familiar with the model representation as it is emphasised in the earlier years of Primary schooling. The goal, however, is to help students transition to the algebraic method. Instead of making the switch of modes abrupt for students, the effort to place the representational modes side-by-side (as shown in Figure 2) in a tightly-linked way is to strengthen the links among them by focusing on the Visual connections. I surmise that a continual use of Visual connections in instructional practice, especially in the earlier stages of learning a new mathematical tool or language (as in the case of Algebra), can potentially help students smooth the intermodal transition and strengthen their mental movement between modes when solving related problems.

3 Idea Connections

By "Idea connections", I have in mind conceptual links established among mathematical knowledge bits. Such connections can be made at various levels of grain-sizes within the mathematical domain. At the smaller grain-size, connections can be considered at the "local" level, such as within a "topic" in teachers' Scheme of Work. An example is the connections among Square, Rectangle, and Parallelogram in the teaching of geometry. If one views these geometrical objects merely as separate entities, one misses out on the richness of relationships and reasoning that are vital processes in the discipline of mathematics. On the contrary, if one sees the inclusion relations among these special quadrilaterals, new opportunities open up to the students. For one, they need not think of the attributes of these special quadrilaterals as separate sets of properties for separate figures; rather, all these quadrilaterals share the common properties of the parallelogram, with additional constraints imposed with each step down the hierarchical tree of special quadrilaterals. Moreover, the reasoning process behind the inclusion relations is in itself a valuable mathematical method: Students learn that what is established for a set applies also for its subset, and the power of this reasoning can be used in other similar mathematical arguments.

Beyond the immediate topic, Idea connections can also be made. Extending from the example on special quadrilaterals taught usually in Year 7 in Singapore, connections between special quadrilaterals can be made at Year 9 to the topic of "Cyclic quadrilaterals". Apart from learning that opposite angles of cyclic quadrilaterals are supplementary, a question can be asked if the converse is true: that quadrilaterals whose opposite angles are supplementary are cyclic. This naturally leads to viewing a cyclic quadrilateral as a special quadrilateral whose place can be determined among the special quadrilateral hierarchical tree.

Idea connections in Secondary Mathematics can go even beyond traditional domains such as geometry. In the case of problems about special quadrilaterals such as the proof that the diagonals of a parallelogram bisect each other, apart from tackling it within the geometrical domain, one can sidestep by making a connection to coordinate geometry, and then using algebraic manipulation based on the coordinates of the vertices to show easily that the midpoint of pairs of opposite vertices of a parallelogram are coincidental, thus proving the conclusion needed. Connections can thus be made between traditional domains of geometry and algebra through coordinate geometry.

4 Temporal Connections

Although a lesson is often taken as a convenient unit of planning and analysis, the consideration about the tools, materials, and sequences of the lesson often go beyond a single lesson. Teachers think about the content suitable for a lesson in terms of what goes before and what is to come after: the lesson builds upon the foundation from the previous lesson(s) and anticipates the trajectory of the "flow" into the proceeding lesson(s). Rephrased in the language of "connections", elements within a lesson not only connect with each other within the lesson; rather, it often also connects with what was taught before as well as anticipates connecting with what is to come in subsequent lessons.

As an example, when we think of teaching the Year 9 topic in Singapore on "angle properties in a circle", we do not think only in terms of isolated lessons in the topic; rather, we think first of the topic as a module, and then in terms of lessons as inter-connected units within the module to help students view theorems in the topic as related to each other. For a start, we will introduce the theorem that (1) "angle at the centre is twice the angle at the circumference subtended by the same arc" in the first lesson. In the second lesson, results (2) "angle in the semicircle is right" and (3) "angles in the same segment are equal" will be introduced. (2) and (3) will not 'stand alone' in isolation but will be taught as straightforward corollary and application of (1). (3) can also be seen as in anticipation of (4) "angles in opposite segments are supplementary" which can be introduced in the third lesson via the question, "what happens if the angles are not in the same segment?" In this way, the language of "same segment" is revised and positioned to lead to the contrast of "opposite segments". As (4) can be easily derived from (3) or (1), these logical links can then be used to strengthen the connection within the elements in the module.

As seen from the above example, Temporal connections are not categorically different from Visual and Idea connections. In fact, we see the former as consisting of the latter categories and threaded through time. In other words, while Visual connections highlight the importance of linking modes of representation and Idea connections feature the relationships among mathematical ideas, Temporal connections focus on another dimension: Time — the need to build Visual and Idea connections consistently over time. Figure 3 illustrates the relationship among these types of connection succinctly. I advocate a view of making mathematical connections not merely as occasional events once in a while, but a regular work of instructional practice carried out in a sustained way over time.

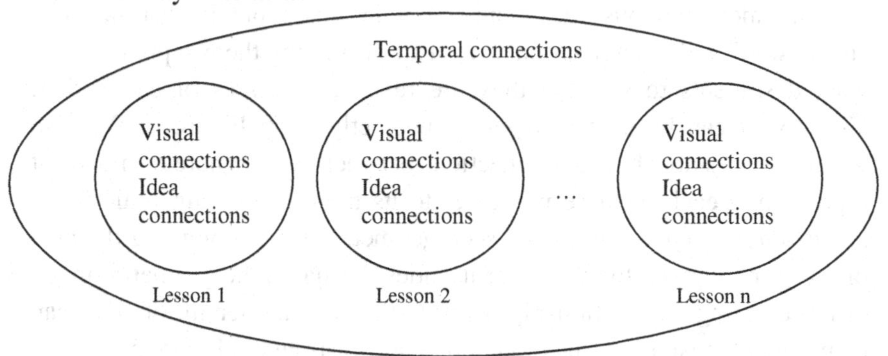

Figure 3. Relationship between the types of mathematical connections

5 Project on Making Connections

We embarked on a project that was based on the constructs of Visual connection, Idea connection, and Temporal connection in the context of teaching a Year 8 class in a Singapore Secondary School. Discussions among teachers prior to the project were over the difficulty that students — particularly students who were mathematically-challenged — in the school faced when confronted with algebraic manipulation. The teachers reported that students made many mistakes in symbol manipulations and it seemed the students could not make sense of the basic rules and symbols of algebra. The goal of the project was thus to design a module that would help students make sense of the algebra they do. Seen through the language and perspective of "mathematical connections", we aimed to help students start with concrete representations of algebra and then connect it gradually to the symbolic form of formal algebra over the course of the module.

We selected students from the Normal (Academic) class to participate in the study. These selected students were judged by their resident teachers to be among those who had the most difficulty with algebra. A total of sixty students attended the module. The topic selected for the module was "expansion and factorisation involving quadratic polynomials" as it was agreed among the teachers to be the most challenging topic for the students.

The most obvious Idea connection in this topic is that between "expansion" and "factorisation". Instead of treating them separately, we wanted students to see that they are reverse processes of each other. Visually, it can be presented more succinctly as in Figure 4. We also wanted to establish the connection between the concrete mode of representing each of these processes to its more formal algebraic form. For example, expansion can be seen geometrically as obtaining the area of a rectangle given the length of its sides. Visually, the geometric mode of representing the relationship between the area and length of sides can be placed alongside the algebraic mode, as illustrated in Figure 5.

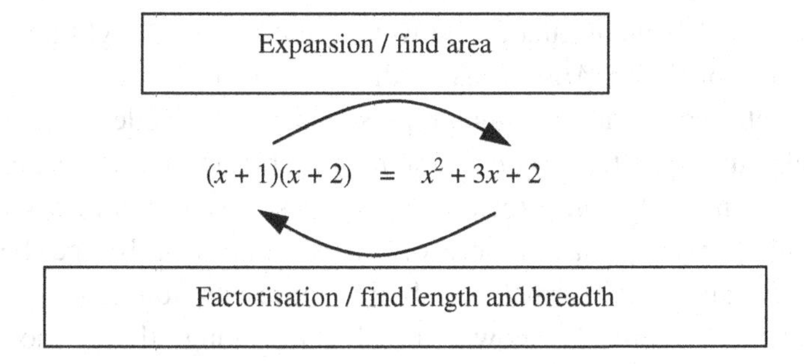

Figure 4. Idea and Visual connections between expansion and factorisation

Factorisation, however, may not be as straightforward to the students as expansion. We stayed with the geometric analogy of factorisation as finding length/breadth of the rectangle given the area. To concretise this "forming of rectangle" stage, we introduced *AlgeCards* for a start. *AlgeCards* are similar to Algebra tiles with two differences: (1) it is school-made laminated vanguards instead of commercially available tiles to save cost; and (2) "x^2", "x", and "1" are imprinted on the cards to help students make clearer visual connections between the geometrical and the symbolic modes.

Expansion	Find area
$(x+1)(x+2)$ $= x^2+2x+x+2$ $= x^2+3x+2$	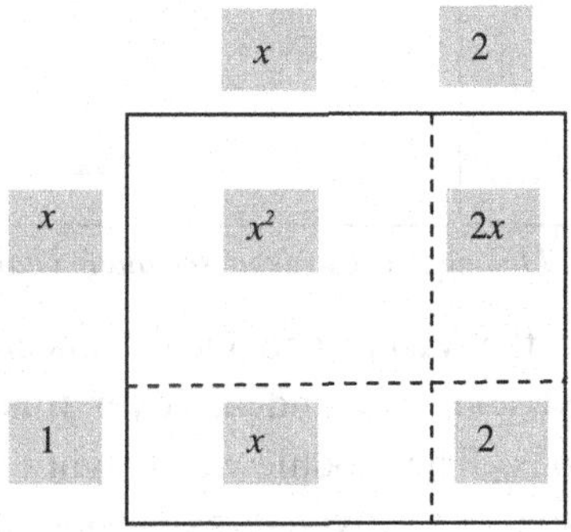 x 2 x: x^2 | $2x$ 1: x | 2

Figure 5. Connecting the geometric and algebraic modes of expansion

The purpose of using the *AlgeCards* is to help students actively carry out the "forming rectangle" as an essential part of factorisation in a concrete way. We were mindful, however, that students ought not to stay too comfortable with *AlgeCards*; rather, we wanted students to make entrance into factorisation using a representation that made sense to them but would subsequently progress to a method that approximates algebraic dexterity. Since "forming rectangle" is the key step, we also included "Rectangle Diagram" as a sort of visual and idea bridge between *AlgeCards* and the formal symbolic representations of quadratic factorisation. Figure 6 shows the links among these modes of representation. "Rectangle Diagram" is a simplification of the *AlgeCards*; and unlike the latter, it can be easily drawn and thus portable as a useable tool in paper-and-pencil contexts, including test situations. Seen through the now well-known construct of "folding back" in Pirie-Kieren's theory of dynamical growth in mathematical understanding (Pirie & Kieren, 1994), the presentation of factorisation in multiple modes facilitates students' folding back to understandings that require concrete materials such as the *AlgeCards*.

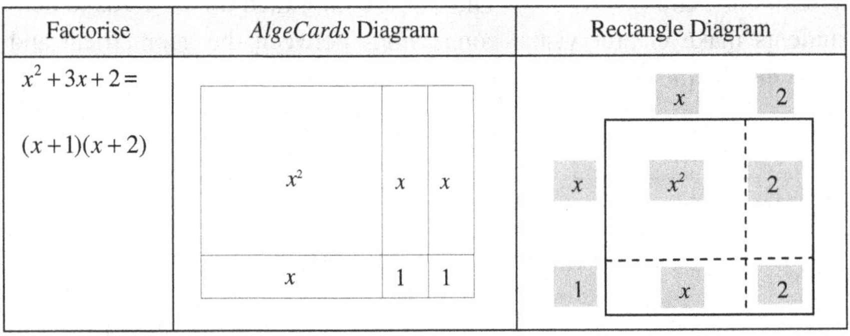

Factorise	*AlgeCards* Diagram	Rectangle Diagram
$x^2+3x+2=$ $(x+1)(x+2)$	x^2 x x x 1 1	x 2 x x^2 2 1 x 2

Figure 6. Linking AlgeCards to Rectangle Diagram and to the algebraic factorisation

Since the Visual and Idea connections discussed above were to be developed with the students over time, we next considered Temporal connections. The module was taught over three one-hour lessons. In the first lesson, we covered expansion and focused on introducing the (Visual and Idea) connection between expansion and the geometric equivalent of finding area given length/breadth of rectangle. To reinforce this connection, students attempted a worksheet that involved numerous expansion tasks represented in a form similar to the one shown in Figure

5. In Lesson 2, factorisation was introduced as the reverse of expansion (with Figure 4 shown in class). We started with the use of *AlgeCards* and students were given a worksheet comprising many simple quadratic polynomials for their practice of the use of the *AlgeCards*. While the focus in Lesson 2 was on familiarity with the *AlgeCards* as a way to make sense of the factorisation as finding length/breadth given area, the Rectangle Diagram and the algebraic mode were also included in the worksheet from the start to help students get used to the intermodal links among these modes of representation. Items in the worksheet during Lesson 2 are similar in form to those shown in Figure 6.

In Lesson 3, we revised at the start of the lesson by making connections back to the emphasis in Lesson 2: that factorisation can be viewed as finding the length/breadth given the area of a rectangle. However, the main focus of Lesson 3 was to take a step away from the *AlgeCards* to transit towards the Rectangle Diagram. In other words, instead of using the actual manipulatives of the *AlgeCards* to form the rectangle, the students were encouraged to use the Rectangle Diagram directly and use paper-and-pencil to check if the component terms they attributed to the respective component rectangles in the diagram matched the original expression. Nevertheless, to help prevent an abrupt switch of representational modes, students were still given the *AlgeCards* so that they can "fall back" on using them should they need a slower transition for themselves. In this way, students can pace for themselves how ready they were to "wean off" from the AlgeCards as a way to perform quadratic factorisation.

6 Project as a Teacher Development Enterprise

Alongside the goal of helping students make mathematical connections in the learning of algebra, this project was also conceived as a platform for teacher development. The teachers involved in the school formed a *Lesson Study* team prior to the start of the project for the purpose of learning from students' work in the context of an instructional innovation. They invited me into the group as a Knowledgeable Other. Guided by the stages advocated by Lewis (2002), and Stepanek, Appel,

Leong, Mangan, and Mitchell (2007), we used the common features of their *Lesson Study* model to guide the entire process: We met to discuss the difficulties students faced and identified the goals of the project; that was followed with more discussion meetings on the design of the module — the details are given in the previous section of this chapter; subsequently one teacher in the team carried out the teaching of the module. All three lessons in the module were treated as research lessons — other teachers in the team sat in for the lessons, made disciplined observations, and shared in post-lesson meetings after each lesson. The teachers who participated in the team shared that they learnt much both from the approach taken in the module as well as the observation of students' work in class. Details of some relevant data and learning points from the perspective of the teachers can be found in Leong et al. (2010).

For the purpose of this chapter, I turned my attention to another target group of this teacher development enterprise — the student teachers (ST) enrolled in the Postgraduate Diploma in Education (Secondary) Programme 2010 cohort Mathematics methods course at the National Institute of Education (NIE) in Singapore. Arising from the positive learning experience shared by teachers in the *Lesson Study* team, I wanted to explore if STs at NIE would also benefit from accessing this instructional innovation.

All the relevant data from the school — videos of classroom happenings, videos of teachers' and students' reflections, lesson plans and worksheets — collected from the *Lesson Study* enterprise were processed for the purpose of packaging them in a way that would enable the STs to follow the goals and flow of the project we carried out in the school. For example, videos were cut into snippets of no more than four minutes surrounding, say, the teacher's demonstration of a particular example of factorisation. The processed data were then re-organised along the structure as shown in Table 1.

The contents were packaged into an e-learning suite. STs accessed the data and information as well as attempted the Milestone Tasks solely through electronic means over a recommended total time — not necessarily continuous in one go — of six hours. The timing of the availability of this e-learning package to the STs is poised between the two mathematics method course semesters, partly for the purpose of connecting these two parts of their learning to teach mathematics; particularly, they can use the learning points from the suite to improve upon the lesson plans they devised in the first semester to prepare for the micro-teaching component in the second semester.

Table 1

Structure of the e-learning suite that student teachers accessed

Section	Main content	Data from Lesson Study project	Post-section milestone task for student teachers
1	Overview of instructional innovation	Lesson plans and worksheets used; article by Leong et al. (2010).	Milestone Task 1: Tasks that check STs' proficiency in the use of the Rectangle Diagram in expansion and factorization
2	Zoom-in to Lesson 2	Selected video snippets of Lesson 2 that highlight the key moves of the lesson, including selected snippets of students' seatwork and discussions.	Milestone Task 2: A series of questions about Lesson 2
3	Zoom out to before and after Lesson 2: Lessons 1 and 3	Selected video snippets of Lessons 1 and 3 that highlight the key moves of the lesson, including selected snippets of students' seatwork and discussions.	Milestone Task 3: A question about STs' overall responses to the module
4	Views of teachers and students	Video snippets of interview sessions conducted with selected teachers in the Lesson Study team. Video snippets of interview sessions with two students selected by the teachers as among the most mathematically-challenged.	Milestone Task 4: A review of STs' responses in Milestone Task 3 in the light of feedback from teachers and students
5	Overall reflections	None	Overall response to the package in relation to ideas about teaching mathematics covered in the Mathematics methods course at NIE

7 Data and Analysis

A total of 130 STs were enrolled in the 2010 Mathematics methods courses. All of them participated in the e-learning suite. For the purpose of this chapter which focuses on mathematical connections, I looked at the more relevant online responses for Milestone Task 2 and Milestone Task 3. The set of questions for Milestone Task 2 is as follows:

1a) What do you think about the whole approach of conveying expansion as finding area given length and breadth of rectangle and factorisation as the "reverse process" of finding length and breadth given area of rectangle? Support your answer by making reference(s) to well-known sound instructional practices/theories.

1b) Is there something you can learn from this approach that can be incorporated into the revision to your lesson plan for the upcoming microteaching?

2a) Discuss the role of AlgeCards in the instructional process in this lesson.

2b) Does the lesson resource that you prepared for microteaching contain some tool/scaffold that plays a role similar to the AlgeCards in this lesson? Why or why not?

3a) How is the smooth transition from AlgeCards to Rectangle Diagram critical to students' learning about quadratic factorisation in this lesson? Make reference(s) to the videos in your answer to this question.

3b) Is there an important transition to be made in your lesson for microteaching? If so, how can you make the transition as smooth as possible for your students' learning?

4a) Is there evidence to suggest that students in the videos benefitted from this mode of learning? If so, what do you think are the main ingredients that contributed to its success?

4b) From your answer to 4(a) above, are there some learning points that you can apply in considering revision to your lesson plan for microteaching?

The question for Milestone Task 3 is:

> Imagine you are a team member in this teaching innovation project and you are about to carry out the module of 3 lessons for your resident Sec 2NA Mathematics class. What are some things you buy-in to the innovation? What are some things you will change/add? In each case, provide supporting reasons by way of appealing to sound practices or theories you have learnt in the last semester.

I further zoomed-in to questions 1(a) and 3(a) of Milestone Task 2 as they relate directly to the focus of making mathematical connections as the theme of this study. The responses to the question of Milestone Task 3 were examined for evidence of student teachers' attempt at observing Temporal connections over the three lessons. For all the responses, I first looked for any references made towards the importance of making connections. Among those STs who made explicit connections they noticed, I examined their responses more closely for emerging notions they expressed about the place and nature of connection-making in their view of mathematics instructional practice. In the following sections, the data reported is based on a preliminary analysis of sixteen STs from one tutorial group in the cohort.

8 Online Responses in Relation to Visual Connections

Seven STs made reference to the CPA (Concrete-Pictorial-Abstract) approach in their responses to Question 1(a) of Milestone Task 2. That CPA was mentioned by a significant number of respondents was especially interesting given that nowhere in the online suite was there an explicit reference to this term. CPA is one instructional approach advocated by some Singapore mathematics educators (e.g., Chua, 2010; Edge, 2006; Wong, 2010) for the purpose of helping students access materials deemed abstract to them by beginning with concrete modes of representation and gradually replacing them with pictorial forms before introducing the abstract representation. This approach of presenting mathematical ideas was covered in the methods course through the mass lectures. To these six STs, they identified the innovation carried out by the *Lesson Study* team as an application of the CPA instructional approach.

Beyond merely stating CPA, a number of these respondents went on under Question 3(a) to elaborate on the criticality of a careful linking of these modes of representation in the learning of quadratic factorisation:

> A smooth transition from AlgeCards to the Rectangle Diagram is crucial for the students' learning... in the following ways: (1) A transition... requires students to realise that the constant term of the quadratic expression... needs to be rectangulated as well and therefore the problem ... can be approached by first considering the integer factors of the constant term... (2) A transition... also requires students to appreciate that the term in x... should be suitably split into a sum of linear terms... [to] fit into the entire rectangle's top-right and bottom-left regions. [Extract from STB's responses]
>
> The smooth transition is critical because it weans students off AlgeCards... So links between the AlgeCards and the Rectangle Diagrams must be made very explicit to the students. [Extract from STE's responses]
>
> The Rectangle Diagram not only moves on to the pictorial aspect but it delicately moves away from the negative lengths and areas and the videos show that the students learn quite well from this approach. [Extract from STG's responses]
>
> The smooth transition from AlgeCards to Rectangle Diagram is especially important as it allows students to solve problems without relying on concrete manipulatives and aids. ... [W]e saw the lesson carry out this transition in 3 stages: introducing the AlgeCards as the main method of solution, giving a systematic approach still using the AlgeCards, then focusing on the Rectangle Diagram with minimal use of AlgeCards at the end. The key was probably the second step, where the systematic approach provided some ideas on why the Rectangle Diagram would work by getting students to consider the factors of the constant term. [Extract from STH's responses]
>
> This transition allows students to understand and accept the abstract approach, by direct comparison with the concrete approach. As seen

on the video, the two approaches are positioned side by side to allow this comparison to take place. [Extract from STI's responses]

It can be seen from the responses that the STs could elaborate on the actual links within the C-P-A modes in ways that were relevant to the context of the video lessons. They not only concurred that careful linking between adjacent modes were important in the teaching of the topic; the explication of the critical features of the method — such as the systematising of the *AlgeCards* procedure by first considering the constant term, and the "delicate" side-stepping of the problem of negative areas — were also in line with the original intents of the project.

The STs' focus on the different modes of representations — concrete, pictorial, and abstract — as well as the need for tight intermodal links in instruction strikes a chord with the notion of Visual connections we elaborated in the earlier sections of this chapter. In other words, seen through the construct of Visual connections, the STs noticed from the online journey that integral to students' learning was a careful sequencing of relevant visual modes and the weaving of tight links between them so that students can transit from one to another smoothly. In the case of STI, these links need to visually standout to the students explicitly. He/she advocated a "side-by-side" positioning of the modes on the whiteboard "to allow this comparison to take place".

9 Online Responses in Relation to Idea Connections

Apart from CPA, some student teachers pointed out other "links" they noticed in the online package in their responses to Question 1(a) of the Milestone Task 2

[B]y linking factorisation and expansion together, clear links are drawn between the two concepts so excessive memorisation is not needed... [Extract from STA's responses].

The method of teaching factorisation by comparing it with expansion ... may help students form more usable schemas and reduce their dependence on recalling a different set of procedure for factorisation. [Extract from STD's responses].

I like the approach. Students are very familiar with finding area using L[ength] and B[readth] and vice versa. This is enga[g]ing their prior knowledge to introducing the concept of expansion and factorisation. [Extract from STK's responses].

The idea of using area also facilitate[s] the students' learning by bridging the gap between expansion and factorisation. By learning factorisation as the "reverse process", that is finding length and breadth given the area, students will be able to find that the two concepts are related and not as separate topics. [Extract from STL's responses].

I think it is an innovative idea to link expansion and factorisation to area of a rectangle because this motivates the learning of the method and concept of expansion and factorisation for the students. [Extract from STO's responses].

These STs noticed Idea connections of two kinds: (1) between the concepts of expansion and factorization — that one is a "reverse process" of the other; and that (2) these concepts of expansion and factorisation were built upon conceptions of area taught to students in earlier schooling years. All of the STs quoted above highlighted this presentation of tight linkage of concepts in instructional work as a commendable practice.

It is interesting to note that while the STs gave positive reviews about establishing idea connections, they provided different motivations, which include reducing the need for memorisation, developing useful schemas, and helping students view mathematics topics as related.

10 Online Responses in Relation to Temporal Connections

Compared to Visual connections and Idea connections, references in the STs' responses to the Milestone Task 3 to Temporal connections are less conspicuous. This may have to do with the phrasing of the questions. While Questions 1(a) and 3(a) of Milestone Task 2 pointed more directly to making connections, the question in Milestone Task 3 did not require STs to focus on making connections over time. The tweaking of the questions to better fit the focus of the inquiry is certainly an area that requires improvement in a follow-up project. Nevertheless, the responses of a few STs (implicitly) revealed some consciousness of time as well as the need to make development links beyond the confines of a single lesson:

> Although quite a bit of time is being spent on getting the students used to the method, I thought that it will allow students to retain what they've learnt for a longer period of time... [Extract from STE's responses].
>
> I like the whole transition of the lesson[s]. Starting from expansion and using of the area to concretise the concept, the videos show the success of this innovation project. The AlgeCards are good as hands-on for the students to play with before moving on the Pictorial Rectangle Diagram and then the Abstract factorisation itself. [Extract from STG's responses].
>
> [T]he use of AlgeCards throughout the 3 [lessons], as well as the side-by-side presentation of the AlgeCards with the Rectangle Diagram in teaching are 2 aspects of the module which I really liked. [Extract from STH's responses].
>
> I particularly like the reference of expansion as finding the expression for the area of a rectangle given the sides, as... [it] complements what happens in the second lesson during factorisation, where we are given the area of the rectangle and told to find its sides. [Extract from STI's responses].
>
> Although extra time and effort are needed to carry out these lessons, it is actually worthwhile as we can see the improvements made by these weaker students. [Extract from STO's responses].

From the responses, it seems that some STs were aware that for certain teaching methods to effectively help the students, they needed to be sustained over time (that is, beyond the temporal boundaries of a single lesson). For example, STH made reference to the use of AlgeCards "throughout the three lessons" as a feature he/she liked about the module. More than merely the need to maintain the use of a method over time, some of the STs mentioned the importance of making connections in the development of the method (such as from expansion to factorisation) from one lesson to the next (e.g., STG and STI).

Not unexpectedly, some were conscious that perhaps more than the usual amount of time was devoted to the development of the representational modes used in the module. Nevertheless, they think that this investment of time can be "worthwhile" if seen against longer term "ret[ention] what they've learnt".

11 Discussion and Conclusion

Although the language of Visual connections, Idea connections, and Temporal connections were not formally introduced to the STs in the methods course, their online responses indicated that a number of them were making implicit references to these forms of connections and noticing their importance to mathematics instructional practices in the classroom. That they were mostly aware of the significance of mathematical connections in teaching is encouraging.

I am further encouraged to see that the STs were not merely referring to the mathematical connections in a cursory way. As seen from the extracts of their responses, a number of them were able to specify to a high degree of precision the actual links that were critical in the context of teaching the topic. In the case of this project focus which was on quadratic factorisation, they were able to point out the fine-grained details of the connections, such as the expansion-factorisation reversal, the gradual fading away of *AlgeCards* as the emphasis shifted to the Rectangle Diagram, the need to maintain the method over a few lessons, and even the whiteboard side-by-side placing of the modes of representations to help students see the links more clearly. To me, this ability to analyse the different potential connections within the topic (and across topics) is crucial to lesson planning that supports this goal of making mathematical connections a regular feature in our classrooms.

In summary, this chapter highlights three types of mathematical connections: Visual connections, Idea connections, and Temporal connections. It attempted to show that these forms of connections can be weaved into our current curricular structure and taught on a regular basis. With careful planning and design, lessons that are crafted with these connections in mind can be helpful to students' learning of challenging mathematical ideas. At the same time, these connections are perhaps simple and obvious enough for novice teachers to notice as critical to quality instructional practices.

Acknowledgement

I would like to thank Yap Sook Fwe for the help rendered in the project and the comments given to improve the earlier draft of this chapter.

References

Bosse, M. J. (2003). The beauty of "an" and "or": Connections within Mathematics for students with learning differences. *Mathematics and Computer Education, 37*(1), 105-114.

Chua, B. L. (2010). *Teaching Algebra II.* Slides of lecture to July 2010 cohort of PGDE(Sec) student teachers taking the Mathematics methods courses.

Edge, D. (2006). Teaching and learning. In Lee, P. Y. (Ed.), *Teaching Primary school mathematics: A resource book* (pp. 29-46). Singapore: McGraw-Hill.

Healy, L., & Hoyles, C. (1999). Visual and symbolic reasoning in mathematics: Making connections with computers? *Mathematical Thinking and Learning, 1*(1), 59-84.

Horton, B. (2000). Making connections between sequences and mathematical models. *Mathematics Teacher, 93*(5), 434-436.

Knuth, E. J. (2000). Understanding and connections between equations and graphs. *Mathematics Teacher, 93*(1), 48-53.

Leong, Y. H. (2006). Three panels presentation. *Maths Buzz* 7(2), 2-4. Singapore: Association of Mathematics Educators.

Leong, Y. H., Yap, S. F., Teo, M. L., Thilagam, S., Karen, I., Quek, E. C., & Tan, K. L. (2010). Concretising factorisation of quadratic expressions. *The Australian Mathematics Teacher, 66*(3), 19-25.

Lewis, C. (2002). *Lesson Study: A handbook of teacher-led instructional improvement.* PA: Research for Better Schools.

Lowrie, T. (2011). Visual and spatial reasoning: The changing form of mathematics representation and communication. In B. Kaur & T.L. Toh (Eds.), Reasoning, communication and connections in mathematics (pp. 149-168). Singapore: World Scientific.

Ministry of Education. (2006). *Mathematics syllabuses - Secondary.* Singapore: Author.

Stepanek, J., Appel, G., Leong, M., Mangan, M. T., & Mitchell, M. (2007). *Leading Lesson Study: A practical guide for teachers and facilitators.* CA: Corwin Press.

Pirie, S. E. B., & Kieren, T. E. (1994). Growth in mathematical understanding: How can we characterise it and how can we represent it? *Educational Studies in Mathematics, 26*(2-3), 165-190.

Wong, K. Y. (2010). *Learning Theories I.* Slides of lecture to July 2010 cohort of PGDE(Sec) student teachers taking the Mathematics methods courses.

Chapter 13

Numeracy: Connecting Mathematics

Barry KISSANE

This chapter discusses the nature of numeracy, which is concerned with the use of mathematics in context, reflecting some recent Australian work concerned with numeracy across the curriculum, and some Australian initiatives to highlight mathematics in wider contexts than the mathematics classroom. Australian national curriculum developments, in which numeracy is identified as a General Capability to be addressed in all school curricula, not only in mathematics curricula, are described. In the primary years, situations that demand some level of mathematical thinking arise in all aspects of the curriculum, sometimes unexpectedly, and thus provide opportunities for numeracy to be developed providing we are alert to the possibilities, while in the secondary years, mathematics is important for learning in other subjects, although this is not always well recognized. Students need to develop a capacity to use mathematics in their everyday lives, their school lives and eventually in the world of work. The chapter explores examples of numeracy demands and opportunities across the curriculum, and considers the roles of teachers to develop the necessary mathematical, contextual and strategic expertise to address these.

1 Introduction

Numeracy is fundamentally about using mathematics. Mathematics enjoys a strong place in schools, in part because of a widespread belief that it is relevant to people's lives outside and after school. Indeed, when asked to give a rationale for school mathematics, the most likely response from parents and the community at large, and even students, is that it is "useful". It is a common view that mathematics is connected with other aspects of people's lives.

In elaborating the aims of mathematics education in Singapore schools, for both primary and secondary students, the Ministry of Education has referred explicitly to these connections. For example, both primary and secondary syllabuses have a common observation:

> Connections refer to the ability to see and make linkages among mathematical ideas, between mathematics and other subjects, and between mathematics and everyday life. This helps students make sense of what they learn in mathematics.
>
> (Ministry of Education, 2006a, p. 6)

In this chapter, emphasis is on the second and third aspects of connections: those that link mathematics to the learning of other subjects and those that link mathematics to activities in everyday life.

In considering the question of mathematics in everyday life, and to understand the associated ideas at a personal level, the reader is invited to consider what mathematics they have used in their "everyday life" during the last week. For this purpose, do not regard yourself as a specialist teacher of mathematics in primary school or secondary school, but focus on other aspects of your life: as a parent or a child; as a consumer or a manufacturer; as a neighbour or a colleague; as a citizen or a tourist; as an employee or as an employer; as a spectator or a shopper; as a client or a patient. To understand the role of mathematics in everyday life, an important first step, although a difficult step, is for mathematics teachers to consider the everyday life that does not include being a teacher of mathematics to children. There are many activities in which Australian adults routinely engage that use mathematics in some ways, as Hogan, Van Wyke and Murcia (2004) noted:

> ... household budgeting; understanding bank statements and fees; paying bills for power, water and rates; running a vehicle; filling in

> application forms; understanding prescription labels; buying appliances; and credit card use. People rent houses, buy houses, build houses or renovate houses. People buy, lease and rent cars. People shop and cook. They design, make and create things. They sew, design gardens and make furniture. They find their way around using street directories. They have hobbies like making clothes, designing a patchwork quilt and painting a landscape. All these activities involve decisions about competing demands of cost, budget and design solutions on the basis of what is necessary, desirable and feasible. ... People can engage with the mathematical demands of such tasks to a greater or lesser extent. Indeed, their perception of the task as being mathematical can be quite varied. Typically, tasks with significant numerical demands are seen as mathematical; those involving spatial knowledge or estimations are more likely to be seen as "common sense". (p. 26)

Perhaps ironically, much of the use of mathematics in these sorts of ways was not developed directly as a result of what was taught in schools, and has a different character from the mathematics that students generally encounter in both textbooks and classrooms. Yet, such uses of mathematics are important to adults, once they have left school, and so the school mathematics curriculum needs to prepare students to address these.

To give a small and personal illustration, the author was recently visiting Singapore. All prices are of course routinely listed in Singapore dollars. To get a better sense of the prices of things, conversion to Australian dollars was desirable, but the exchange rate at the time (one Australian dollar was then equivalent to about $1.30 Singapore dollars) made this slightly inconvenient. Division by 1.3 is awkward (mentally), and rarely was it important enough to bother using a calculator (which was always available on a smart phone). The solution chosen was to multiply Singapore prices by 0.8 (the approximate reciprocal of 1.3), much more easily accomplished mentally. So, for example, a shirt sold for 40 Singapore dollars would cost about 32 Australian dollars. While this conversion is inaccurate, it was quite adequate in the context, always of importance when numeracy is being considered. This example also illustrates that, although much of the arithmetic today's adults learned at school was precise and written, much of the arithmetic they use in everyday life is approximate and mental.

2 What is Numeracy?

The concept of "numeracy" has been discussed and described for around fifty years now, although common agreement on its meaning is still elusive, and there continues to be a variety of terms used to describe it. The concept of "mathematical literacy", as used in the context of PISA (Organisation for Economic Co-operation and Development, 2003) is closely related; perhaps a distinction is that the emphasis in PISA is on assessing the outcomes of the mathematics curriculum, while numeracy is intended to reflect a wider view. Similarly, the term "quantitative literacy" has many connections with numeracy (Steen, 2001), although the apparent emphasis on number reflected in the idea of "quantitative" might be seen as a little restricting. "Numeracy" appears in some official syllabus documents in Singapore, although very rarely. For example, in the opening paragraph of the rationale for secondary school mathematics, the Ministry of Education (2006a, p. 1) notes: "Students develop numeracy, reasoning, thinking skills, and problem solving skills through the learning and application of mathematics."

In contrast, "numeracy" is commonly used at many levels in Australian discussions about the school curriculum. A useful starting point for the present chapter is that suggested by the Australian Association of Mathematics Teachers (1997, p. 15): "To be numerate is to use mathematics effectively to meet the general demands of life at home, in paid work, and for participation in community and civic life."

In school education, numeracy is a fundamental component of learning, discourse and critique across all areas of the curriculum. It involves the disposition to use, in context, a combination of:

- underpinning mathematical concepts and skills from across the discipline (numerical, spatial, graphical, statistical and algebraic);
- mathematical thinking and strategies;

- general thinking skills; and
- grounded appreciation of context.

This starting point reflects some key characteristics of numeracy. It is concerned with using the mathematics that has already been learned, which in turn involves the user making decisions regarding which aspects of mathematics are likely to be helpful in a given situation. It is concerned with use in a context, including in particular contexts that are outside school (especially, but not only, for the case of adults). Viewed in this light, numeracy would seem to be of direct importance to a discussion of "connections" between mathematics and other activities.

2.1 *Australian Association of Mathematics Teachers' perspectives*

The Australian Association of Mathematics Teachers (AAMT) is the peak professional group of mathematics teachers in Australia, and has contributed significantly to Australian thinking and school practices regarding numeracy. As a voluntary and independent professional association, like the Association of Mathematics Educators (AME) in Singapore, it has provided a mechanism for professional mathematics teachers, educators, researchers and scholars to collaborate, although it does not have direct responsibility for school curriculum or policy.

A number of perspectives about numeracy have emerged from such collaborations, resulting in the positions noted in the preceding section. These include the recognition that literacy and numeracy are distinct attributes; while each is important for school children, and there are similarities between them, there are important differences as well. Teachers need to consciously plan for the development of numeracy, and cannot expect it to be a natural by-product of literacy.

Despite the first three letters of "numeracy", it is important that numeracy is recognised as involving more mathematics than merely "number", and so is substantially different from what is sometimes described as "number sense". While many numerate behaviours are concerned with numbers in some way, there are many others that are not, in the same way that there are many aspects of mathematics that do not directly involve numbers. Numeracy involves much more than mere arithmetic.

Numeracy is sometimes regarded, inappropriately, as a synonym for school mathematics. Such a view can reflect both a limited view of mathematics (e.g., as concerned mostly with computation) and also a limited view of numeracy. While school mathematics ought to contribute to the development of students' numeracy, it has a number of other purposes as well. Similarly, in addition to a command of mathematics, numerate behaviour requires various other kinds of expertise.

In a school setting, numeracy is inherently cross-curricular, since the opportunities and expectations for students to use their mathematics effectively occur in other areas of the curriculum, such as science, health, social studies and sport. A consequence of this perspective is that numeracy is not the responsibility of mathematics teachers and the mathematics syllabus alone, but rather requires efforts from elsewhere in the school curriculum and elsewhere within the school. In this sense, attention to numeracy is important for teachers of all school subjects, in precisely the same way that attention to literacy is the concern of all teachers, not only teachers of students' mother tongues.

Numeracy is always related to a particular context, so that numerate behaviour depends on both knowledge of mathematics and also knowledge of the context. In the example of currency translation above, the context did not require a precise result, but merely a good approximation. Getting a quick sense of cost was more important than a result that was accurate to the nearest cent. Mental methods were appropriate in that context. Yet other contexts may require a more careful, more precise process.

Finally, while distinct from school mathematics, numeracy requires school mathematics to be taught and learned well. Frequently, a deep knowledge of the mathematical ideas involved is needed in order to make sound decisions in everyday settings. This may involve students in inventing their own procedures or using mathematical knowledge in personally new ways; such activity is unlikely if students have learned only standard algorithms, well-rehearsed procedures or remembered routines for undertaking mathematical tasks. In short, students are restricted to using mathematical ideas that they already know well, and have probably learned in previous years.

2.2 *A Framework for numeracy*

Over the last decade, Australian researchers John Hogan and Sue Willis developed a framework for considering the development of numeracy, recognising that different kinds of knowledge (mathematical, contextual and strategic) were to be developed and that learners need at various stages to develop and adopt different roles (fluent operator, learner and critic). Table 1 shows this framework, as represented in a brochure developed as a result of a school-based research project (Western Australian Department of Education and Training, et al., 2004):

Table 1

A numeracy framework (Hogan and Willis)

Being (becoming) numerate involves a blend of three types of know-how:	
Mathematical	Knowing (or learning) the concepts, procedures and skills which comprise the content of school mathematics.
Contextual	Knowing (or learning) the meaning and sense of mathematical terms and processes as used within particular contexts
Strategic	Having (or developing) the orientations and strategies to manage one's way through routine or non-routine problem situations
Being (becoming) numerate involves being able (learning to) to take on three roles:	
The fluent operator	Being (or becoming) a fluent user of mathematics in familiar settings.
The learner	Having (or developing) a capacity for the deliberate use of mathematics to learn.
The critic	Having (or developing) a capacity to be critical of the mathematics chosen and used.

The brochure from which Table 1 was extracted is available for downloading and includes several good examples of both the different kinds of knowledge involved and the different roles that students need to undertake in using that knowledge effectively. The project noted that, in practice, while classroom observations and teacher reflections gave ample evidence of the importance of the three kinds of knowledge, there was much more limited evidence of teachers highlighting and developing the three different numeracy roles identified in the framework (Hogan, Van Wyke & Murcia, 2004, p. xiv).

3 Numeracy Across the Curriculum

These perspectives on numeracy suggest that consideration of numeracy across the curriculum might be an appropriate mechanism for helping students become more numerate, rather than focussing entirely on the mathematics curriculum itself. In this section, three different Australian activities of this kind are described briefly, to elaborate different possible approaches to this.

3.1 *Research projects*

A research project, *Numeracy Across the Curriculum*, based at Murdoch University involved primary school teachers of older children (Hogan, Van Wyke, & Murcia, 2004). The project made considerable use of the framework proposed in Table 1. A key feature was the identification of "numeracy moments" by teachers, and consideration of suitable responses to these. A numeracy moment was regarded as an incident that occurred in the normal course of events in a classroom that alerted the teacher to the need for students to exercise numeracy in some way, or that indicated that there was a problem of some kind for them to deal with a situation involving mathematics, or that indicated an opportunity for students to develop their numeracy capabilities. The project focused on those classes for which mathematics was not the main purpose (although the teachers taught mathematics to the same students at other times).

Numeracy moments were identified in all curriculum areas, especially as teachers became more familiar with the idea and the need to be attentive to notice the demands and opportunities related to numeracy in the normal course of events. Initially, teachers were not aware of many of the numeracy demands and opportunities in their classrooms, although this improved as they gained experience of identifying numeracy moments. Examples of numeracy demands included the following (from Hogan, Van Wyke & Murcia, 2004):

- In a technology class, students were designing and making a Go-cart (a trolley for going downhill). This required them to make drawings that were suitably useful, consider matters of scale, make suitable measurements at a chosen level of accuracy, estimate material needs, understand the relationships between circumference and diameter of circles (in relation to the Go-cart wheels).
- To interpret a reading related to sheep farming in Australia, students needed to understand that small amounts (such as a quarter) of large quantities (such as 56 million) can result in large quantities. This provided an opportunity to develop a better understanding of very large numbers, well beyond their normal everyday use.
- In relation to a class excursion, students needed to make a realistic drawing of a jetty and understand some of the mathematical ideas of perspective, including the relative sizes of near and far features of the jetty. They needed to deal with issues of scale and proportion, make suitable measurements and understand the significance of inaccuracy in drawings or measurements.
- As part of a study of ancient Egypt, students were asked to represent the flood levels of the Nile River on a graph. This required various aspects of graphing, including choices of scale to suit the page, measurements, alternatives for axes and the use of various units. Students needed to interpret their graph in the context of the study of Egypt.

The project worked with teachers to identify suitable responses to numeracy moments of these kinds. In some cases, these involved explicit attention to the numeracy involved in tasks, helping students to use their mathematics more effectively. In other cases, teachers planned later classroom work as a response to the observations about numeracy demands, including learning particular aspects of mathematics that were involved, but not sufficiently well understood. In still other cases, teachers tried to anticipate numeracy demands of classroom tasks, and to prepare students to meet them. It was clear that the development of student numeracy required teachers to be attentive to both demands and opportunities presented by the school curriculum, but that it was unusual for this to take place naturally. Practical advice for teachers on supporting the development of student numeracy as a result of the project is contained in the associated brochure (Western Australian Department of Education and Training, et al., 2004).

Prior to this study, a comprehensive Australian literature survey of the place of numeracy in school mathematics, together with helpful advice for teachers related to planning in order to attend adequately to numeracy demands in the curriculum was provided by Kemp and Hogan (2000). Readers will find many examples of numeracy across the curriculum in the report, together with substantial advice on ways of responding to the associated needs.

A key characteristic of this view of numeracy is that the idea of responding to authentic situations as they arise is fundamental. This is necessarily different from the normal practice in mathematics classrooms of purposefully constructing situations or tasks for students to address. An important conclusion concerns the need for teachers to be alert to both the need and the opportunity for numeracy to be attended to, even though the class attention is directed at something else. From the perspective of the framework identified in Table 1, while the knowledge of mathematics and of context is essential, so too is the strategic knowledge that poses questions such as: "Will mathematics help here? How? How accurate do we need to be? Does our mathematical result make sense in a context?"

The roles identified in Table 1 need also to be developed in response to situations as they arise. A fluent operator does not rely on a teacher or someone else to direct them to draw upon their mathematical knowledge, but does so independently and confidently. The learner role involves recognising the need to know more about mathematics and its uses for a particular purpose, while the role of critic does not take for granted that proposed or implemented uses of mathematics are appropriate, but delves deeper to decide.

3.2 *Numeracy and families*

A different kind of project, called *Numeracy: Families Working it Out Together: The Opportunities are Everywhere*, was also based at Murdoch University (Department of Education, Science and Training, 2003). This national project took the view that many of the opportunities and demands for numeracy for Australian primary school children occurred outside school, when children were under the care of their parents rather than their teachers. Nonetheless, there are many situations outside school in which students need to connect their learning of mathematics to their everyday lives. The major product of the work was a set of three brochures produced for parents, focusing on children in the early, middle and later years of primary school. These brochures can be downloaded from the reference provided, including versions in languages other than English.

The brochures offered advice to parents to help them identify some of the daily activities that required their children to make use of the mathematics that they had learned in school, or which might provide a stimulus to further mathematics learning. Broad contexts of numeracy for travelling, at home, and for shopping were used as well as a context of sport for the two older groups. While the examples on the brochures are of course Australian in character, many of these readily transcend cultural boundaries and would have relevance in Singapore as well. Table 2 provides some examples of activities highlighted in this way.

Advice of this kind on the brochures partly took the form of suggested "fridge lists," which were intended to be used to stimulate family discussion around numeracy in the everyday worlds of parents and children. The term "fridge list" was used to reflect the Australian household custom of using a family refrigerator as a means of communication, with notices and reminders of various kinds placed upon it. In this case, the advice offered was intended to remind all family participants of the many ways in which mathematics appeared when thinking about family activities.

Table 2

Family numeracy examples (Department of Education, Science and Training, 2003)

Age group	Examples of numeracy suggestions
Early primary years	[At home] When talking about TV programs ask: *What is the time? What time does the program start? Do we have enough time to read this book before it begins?*
Middle primary years	[Sport] Talk about times and records for major sporting events. Ask: *How fast did they swim? Is it faster than last time?* [Shopping] When shopping with children ask questions like: *Have you enough money to buy that? Will you get any change? How much? How much more will you need to buy another one? What can you buy for $2.00?*
Upper primary years	[Travelling] When planning a long car trip involve your child in making decisions about the best way to get there. Ask: *Which is the quickest way? How long do you think it will take?* Ask them to use a map to give you directions. [Shopping] Use shopping specials to talk about items they want to buy. Ask: *What price will it be after the discount? Is it cheaper to buy the large one or two small ones? Is the smaller one half the price?*

In addition to exemplars in particular contexts, the brochures offered advice on the use of calculators with children, based on the research that these are potentially useful sources of opportunities to learn about numbers and also more general questions, which frequently served to stimulate discussions regarding numeracy. Some examples of these questions are shown in Figure 1:

Encourage your child to ask questions like these to help them make sense of their everyday situations...

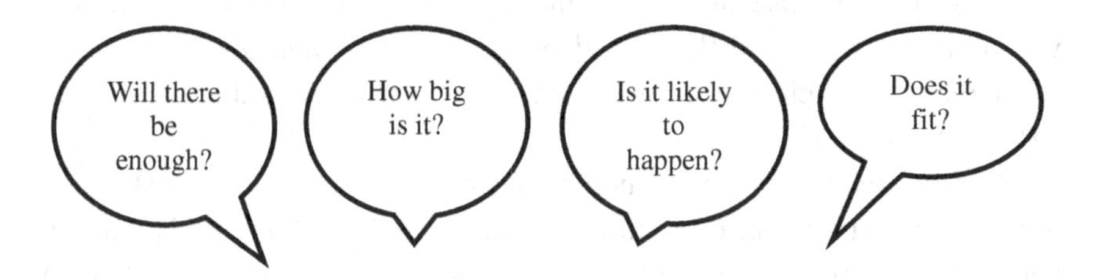

Figure 1. Family numeracy questions
(Department of Education, Science and Training, 2003)

While there are cultural differences, which make direct use of materials from one country potentially problematic in another, the idea of including parents in conversations about the place of numeracy in everyday life, as well as providing some advice on how to think about this question, appears to have considerable merit.

3.3 *Curriculum structure*

A third Australian initiative related to numeracy concerns the official curriculum. Recently, the Australian Curriculum, Assessment and Reporting Authority (ACARA) has developed curriculum for national implementation in a number of learning areas. (In the past, Australia has had separate curriculum structures in each of eight various states and territories, rather than a single national curriculum.) The first curriculum documents prepared and agreed to were for Mathematics, English, Science and History for school years up to Year 10. These have now been accepted by all states and territories for implementation across Australia in the next few years (Australian Curriculum, Assessment and Reporting Authority, 2011a).

Of particular interest to the present discussion is the explicit inclusion of "Numeracy" as one of seven General Capabilities identified as important for Australian students and hence requiring some attention in each learning area. The ACARA documents identify the role of mathematics in the everyday world and its relevance to students: "In the Australian Curriculum students become numerate as they develop the capacity to recognise and understand the role of mathematics in the world around them and the confidence, willingness and ability to apply mathematics to their lives in ways that are constructive and meaningful." (ACARA, 2011b).

Explicit recognition of numeracy in this way has not previously been universal in Australian curriculum domains, and nor has the distinction between mathematics itself and numeracy (concerned in part with mathematics in use) been drawn in such a way consistently. In doing so, ACARA has provided a clear rationale:

> The complexity of contemporary society requires young people to be increasingly numerate. They need to recognise the mathematical basis of authentic problems and engage constructively in their solution.
> The identification of mathematical demands in learning areas enables students to:
> - transfer their mathematical knowledge and skills to problem solving in those learning areas,
> - recognise the interconnected nature of mathematical knowledge, learning areas and the wider world, and
> - become confident and willing users of mathematics in their lives
>
> ACARA (2011b)

Some aspects of mathematics are recognised by ACARA as particularly important for numeracy across a range of areas and in students' everyday lives. These include calculation and number, patterns and relationships, estimation and error, proportional reasoning, spatial reasoning, visual representation, statistical literacy and measurement. However, in addition to such general advice, curriculum documents in individual learning areas highlight aspects of mathematics of particular relevance, in order to assist teachers to be aware of them and to accommodate them appropriately, in supporting the development of student numeracy.

For example, in the English curriculum,

> Students select and apply numerical, measurement, spatial, graphical, statistical and algebraic concepts and skills to real-world situations and problems when they comprehend information from a range of sources and offer their ideas. When responding to or creating texts that present issues or arguments based on data, students identify, analyse and synthesise numerical information and discuss the credibility of sources and methodology.
>
> In the science curriculum,
>
> These include practical measurement and the collection, representation and interpretation of data from investigations. Students are introduced to measurement using informal units in the early years, then formal units; later they consider issues of uncertainty and reliability in measurement.
>
> In the history curriculum,
>
> Students need to organise and interpret historical events and developments and this may require analyses of data to make meaning of the past, for example to understand cause and effect, and continuity and change. This requires skills in numeracy such as the ability to represent and interpret quantitative data.
>
> (ACARA, 2011b)

Both this and the previous two examples of Australian work related to numeracy highlight the important point that, while mathematics teachers are important for the development of numeracy in students, other people have important roles to play as well. Primary teachers are also concerned with numeracy when they are not directly teaching mathematics, secondary teachers of subjects other than mathematics have a role to play, and even parents might be significant agents for developing student numeracy. In other words, mathematics teachers need to collaborate with others in developing numeracy.

The relationship between numeracy and other subjects is not uni-directional, however. Students need to develop numeracy for the purpose of sound learning elsewhere in the curriculum. To succeed in the study of other school subjects, students need to make good use of the mathematics that they have learned. Indeed, the everyday world of school students is largely concerned with school, in contrast to the everyday world of adults, which is largely concerned with work and home. In the next section, we look briefly at the Singapore curriculum to see examples of these connections.

4 Examples of Connections in Primary Syllabuses

Space precludes a complete treatment of this topic, but an inspection of various primary syllabuses makes clear the connections with numeracy. For example, in the upper primary Healthy Diet topic described in the Health Education syllabus (Ministry of Education, 2007) the following are noted as part of the theme of "Knowing what food does for me":

> "vi. Choosing healthier alternatives such as foods that are high in fibre and low in salt, sugar and fat.
> vii. Understanding one's own energy requirements is also helpful to enable one to manage it." (p. 9)

Exploring these themes requires students to deal with the ways in which foods are described in practice regarding these attributes (e.g. on food labels and in charts and tables), as well as to understand how energy requirements are represented, accessed, understood and met. Inevitably, the comparisons involved will require students to draw upon various aspects of measurement, proportions and percentages as well as how to balance these various aspects in practical ways. In terms of the numeracy framework offered in Table 1, it is clear that this kind of work involves both a knowledge of mathematical concepts and procedures, as well as contextual knowledge (e.g., concerning how information about food is communicated via labels and other sources).

The primary science syllabus in Singapore similarly refers to many aspects that require numeracy for successful engagement. Thus, in the domain of Science in everyday life, the following expectations are noted (Ministry of Education, 2008): "Using scientific skills in everyday life, e.g. observing trends and patterns, analysing data from media reports, etc." (p. 3)

Activity of these kinds clearly relies upon statistical skills developed in mathematics, as well as providing opportunities to refine the learning from mathematics in the fresh context of scientific data. The critical role referred to in Table 1 is of key importance in dealing with data from media reports.

The social studies curriculum also offers demands and opportunities for numeracy. To illustrate some of these, the headings below have been taken from Unit P4B: Our Heritage and P6A: Our progress (Ministry of Education, 2005a): "P4B: Our heritage areas, Our multiracial society" (p. 14). "P6A: Concepts of Progress, Globalisation, Industry" (p. 18).

Descriptions of these various attributes require measurements of some kind; heritage areas require some sorts of visual representations; progress is described using trends, with data or graphs; globalisation is a relative term, requiring numerical representation of some sort, and so on. There are clearly many opportunities and expectations here related to numeracy for learning, and sound learning will require that these be somehow taken into account. In terms of the framework suggested in Table 1, strategic knowledge is involved in deciding how, and for what purposes, important social data related to Singapore can be represented helpfully using mathematics associated with indices of society, maps to show distributions, graphs to show trends, and so on.

Numeracy for learning in other parts of the curriculum requires teachers to make some connections of these kinds, even if they are not explicitly written into syllabuses. This is much easier for primary teachers than for secondary teachers, as it is usually part of the primary teacher's responsibility to teach in several areas, and there is thus an expectation that all relevant syllabuses are understood. As with the numeracy project described earlier (Hogan, Van Wyke & Murcia, 2004), teachers can be alert to numeracy moments in areas outside mathematics and help students to focus attention on the "real" connections expected of them. In addition, sometimes primary teachers have more control over time allocations than secondary teachers, who are unavoidably focused on a smaller number of areas of school.

5 Examples of Connections in Secondary Syllabuses

The author conducted a search for the word "numeracy" in various Singapore secondary syllabuses, with no positive results. This observation might be misinterpreted as suggesting both that numeracy was not important for learning in other areas and that other areas do not contribute much to the development of student numeracy. Instead, it is more likely to represent a deliberate choice of curriculum writers to allow teachers to interpret syllabuses for themselves. Indeed, a close inspection of secondary syllabuses makes clear that there are many connections with numeracy.

For example, "Theme 2: Growth of Singapore" in the Secondary Social Studies Syllabus (Ministry of Education, 2005b, p. 8) requires students to learn about the changing population of Singapore and some of the consequences of the nature and rate of population change. It seems inevitable that studying such material will require students to access real available data regarding these matters, such as the two graphs shown in Figure 2, published online by the Singapore Department of Statistics.

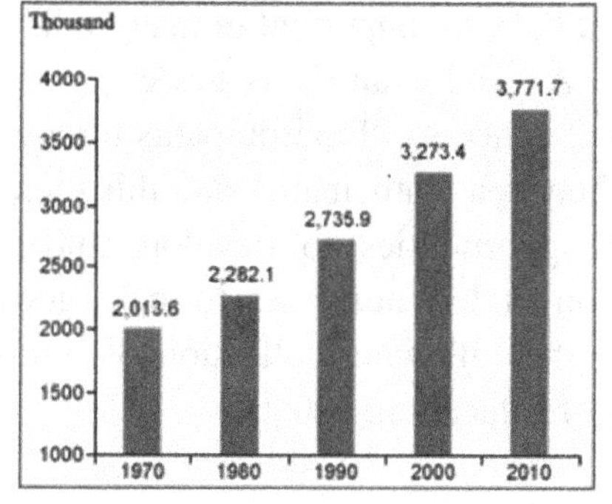

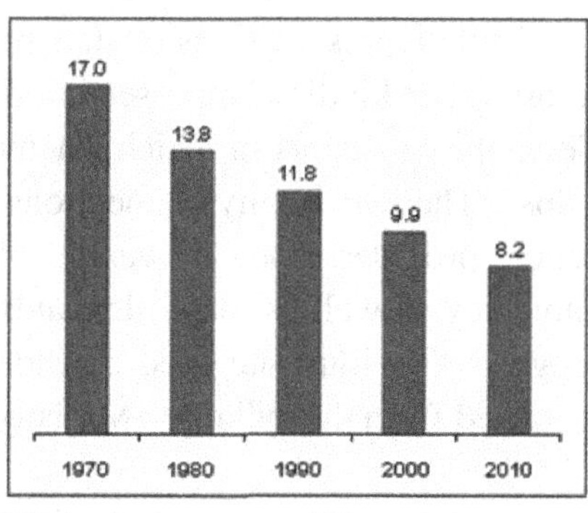

Figure 2. Graphs related to population growth from Singapore Department of Statistics website (2011), used with permission

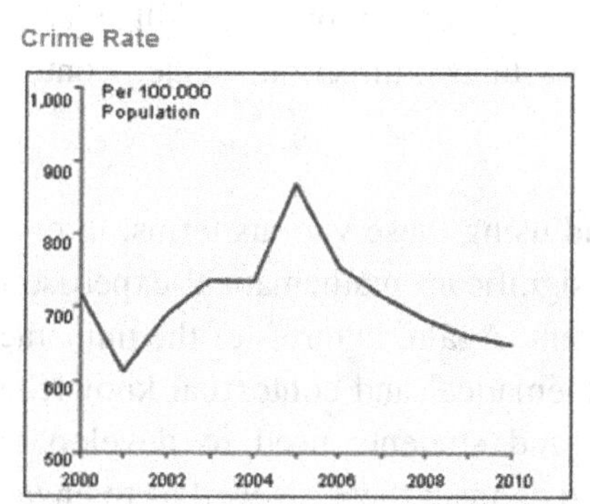

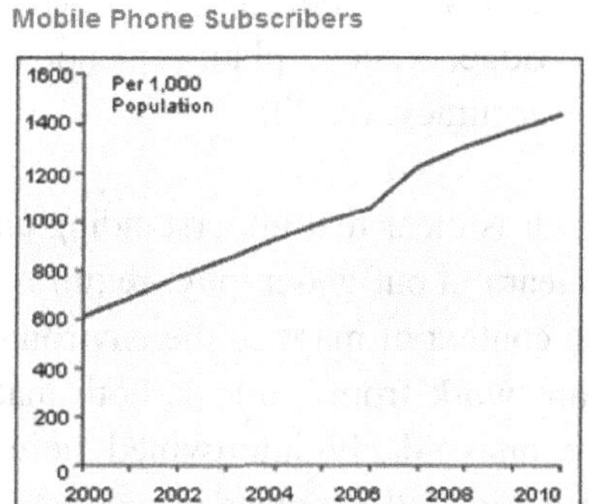

Figure 3. Graphs related to Singapore's people from Singapore Department of Statistics website (2011), used with permission

Similarly, the study of society generally in Social Studies would seem to require students to exercise numeracy to interpret (and, indeed, critique) graphs such as those in Figure 3. Thus, in terms of the framework suggested in Table 1, students need to develop a critical role, which in this case might question the vertical scale chosen to portray the crime rate or discuss the significance of a mobile phone rate in excess of 1000 per 1000 people. Interpretation of graphs and tables is a key aspect of numeracy for learning. There is an excellent chapter devoted to developing expertise with this in the 2010 Yearbook (Kemp, 2010), based on a Five Step Framework to provide systematic advice and specific help.

While representations of statistical data are important in many fields, so are other kinds of representations. A good example is evident in the Geography syllabus in which Theme 2 concerns "Environments through maps". There are many connections between mathematics and this topic, which provides both a wealth of opportunities to develop student numeracy as well as many demands on student numeracy to make good progress. To illustrate this, consider the following collection of terms extracted from the syllabus (Ministry of Education, 2006b):

> Latitude and longitude, physical features, topographical map location (four-figure, six-figure grid references, straight line distances, direction (compass points, compass bearings), line and statement scales, contours (gentle slope, steep slope, mountain, hill, plateau, ridge, valley, plain), location, distance, direction, scale, contour, accuracy. (p. 20)

It is clear that understanding and using these various terms, in order to learn about geography, requires significant mathematical expertise in the context of maps of the environment. Again, in terms of the numeracy framework from Table 1, both mathematical and contextual knowledge are unavoidably intertwined here, and students need to develop the strategic skill to consider how best to connect these, as well as to engage in learning. In this, and many other similar cases, engaging in this process is likely to support the development of student numeracy as well as learning in the content area.

As for the primary years, numeracy for learning in other parts of the curriculum requires teachers to make some connections, even if these are not yet written into syllabuses. A difference between primary and secondary teachers is that it is less likely that mathematics teachers will read other syllabuses unless they do so deliberately: it is not a normal expectation of their work. Yet, to do so might suggest valuable ways for the teacher to help students see the significance of mathematics in other learning areas.

In addition to reading other secondary syllabuses, mathematics teachers may gain a great deal through conversations with specialists in other areas, allowing for a strong focus on the "real" connections expected of their students, with perhaps a little less emphasis on the often sanitised versions of the real world that are seen in many textbooks.

6 Connecting Mathematics Outside the School

While the school curriculum, and hence the work of teachers, is largely responsible for developing important outcomes such as numeracy, it is worthwhile noting that professional and organised activities outside the direct work of schools can also play a significant part. Two current Australian examples, information about each of which is accessible to teachers in Singapore, are described briefly here, to illustrate such possibilities.

6.1 *Reach for the Stars!*

Partly as a consequence of the Australian work earlier referred to, the Australian government has for several years designated a week of the school year as National Literacy and Numeracy Week. As part of this event, since 2003 the Australian Association of Mathematics Teachers (AAMT) has been conducting an activity for students called *Reach for the Stars!* This voluntary event provides an opportunity for mathematics to be connected outside the classroom and even outside the school for those schools and their students who choose to participate.

The essential idea is that classes of school students, with direction from their teachers, collect data locally on some issue that has been chosen as the theme for the year. The data are then submitted to the AAMT for compilation and some analysis, and returned to the respective schools within the confines of the week. This process allows students to use mathematics to explore, represent, compare and summarise an aspect of the everyday world of their school.

Over the years since its inception, many thousands of students from across the country have collected and used data of many kinds as part of the project. These data have included the heights of students (boys and girls, in different year levels), travelling to school (distance, time, method of travel), money trails (using collections of coins placed on the ground, focusing on lengths, values and the ages of 20-cent coins in Australia), height and foot sizes (considering the relationships between these), throwing a coin at a target (considering accuracy in relation to age, throwing distance and other factors) and student names (particularly the lengths and number of letters in students' first names). For all of these, students can use their mathematics, at their own level of sophistication, to compare their class with those of other schools and the national averages, as well as considering differences between observations related to different states, genders, ages, etc.

The most recent example (September 2011) involved the theme of "What does *your* classroom look like?" Questions such as "How big?" "What colour?" "How many?" "How do you know?" were asked and answered. Students were involved in counting, grouping, categorising, ordering, measuring, modelling, calculating, comparing, contrasting, predicting, and explaining matters of interest under the broad theme. Teachers were given specific advice on activities related to the theme that suited a variety of year levels (early childhood, lower primary, upper primary and lower secondary); these ranged from simple counting and direct comparisons through to more mathematically sophisticated tasks involving complex modelling and problem solving. Suggested stimulus questions included the following, among many others:

- How are we the same? How are we different? What statistical tools should we select?
- How big is your classroom? How many people could fit into it if the classroom was empty?
- What is on the teacher's desk?
- How many pens, pencils, text as and other writing implements are in your classroom? Who has the most in their pencil case?

(Australian Government, 2011)

The national summaries provided various interesting information for students as well as their teachers, to continue to see how mathematics was instrumental in understanding their classroom and its inhabitants in relation to others, as well as learning more about the diversity in their own country. Summaries included a variety of graphs and tables, as well as statistical information such as the following:

- If all 3.5 million of Australia's students lined up one arm's length apart then they would reach from Melbourne to Darwin, which is over 3000 km!
- On average the age of an Australian classroom was 29 years.
- The average "people" perimeter of an upper primary classroom was 37 and the average number of people who would fit in the classroom when empty was about 700. [People units use students instead of metres for measurement.]

In addition, summaries encouraged students to take a careful view of data collected in this way, being alert to missing data, possible errors or misunderstandings and recognising the necessity to "clean" data of such problems. Similarly, "people" units are different for very young children and secondary school students. Interpreting data that have been collected in this way and using it to address various purposes of interest to children themselves is a powerful way of improving various aspects of numeracy.

While this event is free for Australian schools to enter, and not available for schools from other countries, information about the event is freely available to all, including teachers from Singapore, via the website. Many of the ideas in this event for connecting mathematics with the everyday worlds of students and their schools are certainly worth considering in Singapore, contributing to a wider view that mathematics helps to interpret and make sense of the world beyond the classroom.

6.2 *Maths and Stats by Email*

Maths and Stats by Email is a free fortnightly newsletter sent via email only to subscribers. While it is less directly concerned with numeracy than the earlier examples, it nonetheless offers a fresh approach to the idea of connections between mathematics and the real world. This publication by Australia's premier scientific research organisation, the Commonwealth Scientific and Industrial Research Organisation (CSIRO) was launched in March 2010 to provide materials about mathematics for school students, particularly those in the middle years of schooling, aged about 9-13 years. The principal aim of the publication has been to communicate that mathematics is making a valuable contribution to the community, is relevant, beautiful, interesting and enjoyable and provides many employment opportunities. In addressing this aim, the newsletter is also a good example of helping students to see connections between mathematics and other activities, including those in everyday life as well as many occupations. Subscription details of the newsletter can be seen online at Commonwealth Scientific, Industrial and Research Organisation (2011), while the current edition can be seen at http://www.csiro.au/helix/mathsbyemail/newsletter/newsletter.html.

Kissane & McConney (2010) have provided detailed evaluative information on the first year of the publication (when it was entitled *Maths by Email*), based on analyses of newsletter contents and especially on two national subscriber surveys. The data provide substantial evidence that the newsletter was meeting its stated aims well, and had appeal to a wider audience than the targeted group. Most issues of the newsletter have a theme, sometimes related to topical or seasonal events, that is concerned with connecting mathematics to some aspect of the real world, so that it might be regarded as a useful resource for helping students make those connections. The theme is used for a short article about mathematics and statistics and their applications in contemporary world, and a short hands-on activity intended for home use is usually constructed with links to the same theme. There is also an online archive of practical activities that have appeared in the newsletter, which might also help both teachers and students. Although the principal audience of the newsletter comprises students around the interface of primary and secondary school, it is clear from the evaluation work that it has appeal for a wider audience, including secondary school students and their teachers.

7 Final Remarks

The concept of numeracy is a strong way of thinking about connections between mathematics and the everyday lives of students in school, and after school. Australian teachers and researchers have amassed considerable experience with this concept over the past two decades, some of which may be helpful for thinking about connections in Singaporean schools.

A key message of this work has been that it is important to consider numeracy in all parts of the curriculum and the school day, rather than concentrate only on the mathematics classroom. Developing numeracy requires both mathematical expertise and contextual knowledge, which are rarely found together entirely within the confines of the mathematics curriculum, either in its official form or in its enacted form. The concept of numeracy across the curriculum is central to making good progress.

It seems optimistic to expect that school mathematics will be sufficient to develop numeracy without explicit attention being paid to it. For primary teachers, this may involve careful scrutiny of both the mathematics syllabus and other syllabuses under their responsibility. Secondary teachers may need to work more closely than is often the case with other specialist teachers to understand the demands and opportunities for numeracy elsewhere in the curriculum. While numeracy is likely to be most clearly developed in school, opportunities to support the development of numeracy outside the school boundaries and in the home, are also worthy of attention.

References

Australian Association of Mathematics Teachers. (1997). *Numeracy = Everyone's Business*. Retrieved October 28, 2011, from http://www.aamt.edu.au/Professional-reading/Numeracy

ACARA (Australian Curriculum, Assessment and Reporting Authority). (2011a). *Australian curriculum*. Retrieved October 30, 2011 from http://www.acara.edu.au/curriculum/curriculum.html

ACARA (Australian Curriculum, Assessment and Reporting Authority). (2011b). *Numeracy*. Retrieved October 30, 2011, from http://www.australiancurriculum.edu.au/GeneralCapabilities/Numeracy

Australian Government. (2011). *National literacy and numeracy week*. Retrieved October 30, 2011, from http://www.literacyandnumeracy.gov.au/

Commonwealth Scientific, Industrial and Research Organisation. (2011). *Maths by email*. Retrieved October 30, 2011, from http://www.csiro.au/mathsbyemail

Department of Education, Science and Training. (2003). *Numeracy: Families working it out together, the opportunities are everywhere*. Retrieved October 30, 2011, from http://www.dest.gov.au/sectors/school_education/publications_resources/literacy_numeracy/numeracy_families_working_it_out_together.htm

Hogan, J., Van Wyke, J. & Murcia, K. (2004). *Numeracy across the curriculum*. Retrieved October 30, 2011, from http://www.dest.gov.au/sectors/school_education/publications_resources/profiles/numeracy_across_curriculum.htm

Kemp, M. (2010). Developing pupils' analysis and interpretation of graphs and tables using a five step framework. In B. Kaur & J. Dindyal (Eds.), *Mathematical applications and modelling: Yearbook 2010*. Association of Mathematics Educators (pp. 199-218). Singapore: World Scientific.

Kemp, M. & Hogan, J. (2000). *Planning for an emphasis on numeracy in the curriculum*. Retrieved October 30, 2011, from http://www.aamt.edu.au

Kissane, B. & McConney A. (2010). *Evaluation of Maths by email: Final report*. Retrieved October 30, 2011, from http://www.eric.ed.gov/PDFS/ED523870.pdf

Ministry of Education. (2005a). *Social studies syllabus - Primary*. Singapore: Author.

Ministry of Education. (2005b). *Social studies syllabus - Lower secondary normal (technical)*. Singapore: Author.

Ministry of Education. (2006a). *Mathematics syllabus — Secondary*. Singapore: Author.

Ministry of Education. (2006b). *Geography syllabus — Lower secondary*. Singapore: Author.

Ministry of Education. (2007) *Health education syllabus for primary level (restricted)*. Singapore: Author.

Ministry of Education. (2008). *Primary science syllabus*. Singapore: Author.

Organisation for Economic Co-operation and Development. (2003). *The PISA 2003 assessment framework — Mathematics, reading, science and problem solving knowledge and skills*. Retrieved November 21, 2011, from http://www.oecd.org/dataoecd/46/14/33694881.pdf

Singapore Department of Statistics. (2011). *Statistics Singapore*. Retrieved October 31, 2011, from http://www.singstat.gov.sg/

Steen, L. A. (2001) Embracing numeracy. In L. A Steen (Ed.) *Mathematics and democracy: the case for quantitative literacy* (pp. 107-116). Retrieved November 21, 2011, from http://www.maa.org/ql/107-116.pdf

Western Australian Department of Education and Training; Catholic Education Office of Western Australia; Association of Independent Schools of Western Australia Inc (2004). *Numeracy demands and opportunities across the curriculum*. Retrieved October 29, 2011, from http://www.dest.gov.au/sectors/school_education/publications_resources/profiles/numeracy_across_curriculum_booklet.htm

Chapter 14

Making Connections Between School Mathematics and the Everyday World: The Example of Health

Marian KEMP

Curriculum documents often refer to the use of mathematics in the everyday world, clearly of importance to them as students in classes other than mathematics and also eventually as adults. Mathematics teachers often find that resources for teaching and learning, such as standard textbooks, do not connect well to the everyday worlds of the students in their classes. There are many opportunities in the classroom to make connections between school mathematics and the everyday world, however, and this chapter illustrates some of these for the particular context of health. Mathematics is connected with official health syllabus expectations, although this is not evident at first glance, and it is also needed for students to adopt healthy lifestyles in their everyday world. A progression of suitable activities is illustrated in this chapter, highlighting some of the ways in which mathematics is needed to understand key aspects of health. These activities include the mathematics associated with describing diets, measuring what is eaten, understanding public health advice about balanced diets, as well as understanding and using nutritional information on packaged foods. For some activities of these kinds, sound use of technology allows mathematics to be connected to health issues more efficiently. Of necessity, teachers will need to adapt these ideas to the mathematics and cultural backgrounds of their students, as well as recognising that connections of mathematics to the everyday world require specific attention to both mathematics and to the everyday world of their students, of which health connections represent only a single example.

1 Introduction

This chapter proposes that learning mathematics in school should involve the integration of authentic activities from students' everyday lives in their learning in the classroom. This would use interesting and motivating activities to develop the students' mathematics; but more importantly they can be used to help to encourage students to engage in "doing" mathematics in their everyday lives. Indeed, the Primary Mathematics Syllabus in Singapore (Ministry of Education (MOE), 2006a, p. 4) states that it is "crucial" that students develop an "appreciation of mathematics as an important and powerful tool in everyday life". This chapter will focus on the primary school classroom but it is relevant to include that the Secondary Mathematics Syllabus in Singapore (MOE, 2006b, p.1) states that "Students develop numeracy, reasoning, thinking skills, and problem solving skills through the learning and application of mathematics. These are valued not only in science and technology, but also in everyday living and in the workplace".

Through engagement with everyday mathematics the students should learn how to make sense of the quantitative aspects of the world and about the different ways in which mathematics is used in print and online media: to inform, to persuade, to manipulate, to support arguments. As adults they will need to make informed decisions in a wide range of areas, based on quantitative evidence (Gal, 2004; Steen, 1997; Watson, 1995). The journey that starts with learning about numbers and travels through to critical thinking about health, political, environmental and other issues is a complex one. This is a journey that needs scaffolding by teachers from early primary school through to the end of secondary school at increasing levels of sophistication, and it is a challenging one for all involved.

Teachers often base their lessons around a textbook that provides a range of examples and activities to help develop students' mathematical concepts and skills in all the strands of the syllabus. The textbooks provide examples and exercises to help both the teachers and students. There are also resource books like those that are available for Singapore teachers (Kaur & Yeap, 2009; Dindyal, 2009) that provide activities aimed at developing students' problem solving, thinking, reasoning and modeling skills. These have been carefully developed using innovative questioning ideas. The word problems are embedded in realistic (or real world) examples for which students need to learn to read and interpret the questions, to understand the language and to interpret what is required. In Dindyal's book for primary school there are very helpful suggestions for teachers concerning the use of story and realistic problems. He highlights the need for teachers to think carefully about the construction of the problems and the contexts. He gives an example of where students cannot just use one of the four operations without thinking about the context: "Ben runs 100 metres in 15 seconds. How long will it take Ben to run 1 km?" Dindyal notes that students need to think about whether someone could sustain the same speed over that distance (2009, p. 36). Kaur and Yeap (2009) also provide guidelines and questions for use in the primary school using story and realistic problems to encourage the students to reason and communicate.

These kinds of resources are valuable for teachers in providing good learning environments for students. For these problems students need to have the literacy skills to read the question and also to think about the context of the questions so as to decide the correct answer. (Zevenbergen, Doyle & Wright, 2004, p. 41). For example the question concerning how many buses are needed for an outing: "315 students, 60 per bus, how many buses?" needs an answer of "6", not "5 remainder 15". These realistic questions do have value in developing mathematical concepts and skills but this chapter suggests that in addition to these kinds of exercises students should engage in the mathematics of their everyday world where they can be the ones to ask their questions, rather than relying on the teacher to do so, as this is what they will need to do in their own lives.

From an early age students can develop an awareness of the mathematics in their everyday world, without complex mathematical concepts and skills. Prompts from teachers and families can foster an interest in the numerical and spatial aspects of the world around them. They learn to count things; they learn about big and small; they learn about shapes and patterns, and they develop some ideas about sharing. When they get to primary school they mainly develop mathematical concepts and skills in the mathematics classroom without explicit use of mathematics anywhere else. However, in primary schools teachers can integrate mathematics with other subject areas across the curriculum and make connections to everyday experiences to enhance students' learning.

This chapter explores how connections can be made between the Primary Health syllabus, the Primary Mathematics syllabus and students' everyday experiences, in order to help students learn mathematics and use their mathematics for learning. The Health syllabus has been chosen because it relates so much to students' lives, and the lives of their families and friends, outside of school. Examples of activities related to food and health are given to illustrate some ways in which students can bring their everyday lives into the classroom and learn from them. These activities involve different levels of mathematics and are just suggestions; from these ideas teachers will be able to develop their own activities suited to their own students.

2 Integrating Mathematics, Health and Everyday Life

The Health syllabus in the primary school indicates that students learn about physical health, emotional and psychological health, and environment and health. As can be seen in Table 1 one of the topics is "Knowing what food does for me: Types of food and diet, healthy food choices" (MOE, 2007, p.9). This context has been chosen for activities that are relevant to primary school students and that require mathematical application to understand properly. It may not be obvious at first that mathematics is involved as there are no numbers in the table but this chapter should illustrate the need to use mathematics, and the potential to improve mathematical skills, in this context. Connections between mathematics, other areas of the curriculum and everyday life could be made.

Table 1

Extract from the Health education syllabus for primary level concerned with healthy eating (MOE, 2007, p.9)

▪ **Theme**	▪ **Learning objectives**	▪ **Scope of content** ▪ **Lower primary**	**Upper primary**
▪ Knowing what food does for me	▪ Students will be able to: ▪ Identify the different types of food for growth and health ▪ Make healthy food choice to obtain and maintain healthy growth	▪ Eating Right ▪ Good habits include having regular meals (breakfast, lunch and dinner), eating in moderation and chewing well. ▪ There are four main groups of food which are present in the Health Diet Pyramid e.g. rice and alternatives, fruit, vegetable, meat and alternatives ▪ The different groups of food have different functions in the body ▪ Eating a variety of foods means to include foods from different groups	▪ Healthy diet ▪ The Healthy Diet Pyramid is a guide to help one make health food choices ▪ The Healthy Diet Pyramid is a useful guide for planning a balanced diet and achieving good eating habits ▪ The Healthy Diet Pyramid states that a healthy diet should consist of more servings of food at the base of the pyramid and less at the top ▪ The food groups in the HDP contain specific nutrients ▪ These nutrients contribute to different functions in the body ▪ Choosing healthier alternatives such as foods that are high in fibre and low in salt, sugar and fat ▪ Understanding one's own energy requirements is also helpful to enable one to manage it. ▪ Improper diet could lead to some common health problems such as obesity, high blood pressure and high blood cholesterol

This section proposes a sequence of activities to illustrate how teachers might integrate the learning and using of mathematics in school mathematics across the curriculum and with the mathematical requirements of everyday life. Teachers can adapt the activities to suit their students' levels of mathematical knowledge, the time available and the particular interests and cultural settings of the students. The activities may be associated with mathematics or health lessons, or both.

2.1 *Activity one: Recording the types of food eaten by the students*

As an introduction to examining the requirements for healthy eating, and discussing what students can do about it, students can productively keep a record of what they are eating over a day or several days. This involves recording all the kinds of food eaten, but not including serving sizes as measurement of these is in Activity three. Prior to their record keeping, group discussion should lead to possible ways of recording. Individual students may choose different methods or they may all decide to do the same. The teacher may guide this discussion. Students may choose to record by numbers of different foods with a tally system and a frequency table, or input the information straight onto a bar graph or pictograph, or other ways of their choosing, depending on their backgrounds. The Data Analysis sections of the Mathematics syllabus for Primary 1 to Primary 6 give a guide to the mathematics relevant to this activity for the different ages of students. For example: students in Primary 1 collect and organize data and make picture graphs; in Primary 2 they add scales; in Primary 3 they construct and interpret bar graphs; in Primary 4 they construct and interpret tables of data; in Primary 5 they learn about averages and in Primary 6 they interpret pie graphs

The next step is to share their records with other students in the class and start to make comparisons. This should include some group and then class discussion about good ways to record information and about how the class may combine their records to enable an overview of the class. They might try several ways to do this and then do some calculations about the percentage of the class who eat particular kinds of foods or the most common types of food that they eat. Working with percentages could be for Primary 5 students.

This may lead to a class survey of likes and dislikes that are recorded in a table and presented in a bar graph, or a pie graph for older students. Students will need to choose the foods that are chosen for the survey, the number of items chosen and the questions asked. Further discussion can focus on what other members of their families eat, including their likes and dislikes. As part of this discussion students can move on to considering what they should be eating for a healthy diet and the implications for some common health problems such as diabetes and high blood pressure. The theme of physical health is in all years from Primary 1 to Primary 6. These may well be topics suited to the upper primary school students but teachers may judge when these can be considered, and in how much depth.

2.2 *Activity two: Examining the Healthy Diet Pyramid*

The Healthy Diet Pyramid, shown in Figure 1, is available on the Internet from the Health Promotion Board webpage (http://www.hpb.gov.sg/) and it has numbers of servings recommended for adults in each of the categories. As noted in Table 1 the Health syllabus requires students to look at the foods that are healthy with reference to the pyramid.

To really understand the nature of the Healthy Diet Pyramid students need to understand the structure of pyramids, as this is essential to the many-to-few servings concept for categories nearer the top of the pyramid. In the introduction to the activity there can be discussion about the nature of a pyramid.

Students can handle some pyramids, construct pyramids from a net, and discuss why a pyramid was chosen and not some other shape. The class can construct a large pyramid and add in examples of the foods in the categories or draw their own triangles to represent the pyramid and include the examples on that. The Mathematics syllabus indicates that Primary 1 students start to examine 2-D and 3-D shapes using concrete models; building up some understanding of shapes and solids. Later on, in Primary 6, they learn to construct solids from nets, including pyramids.

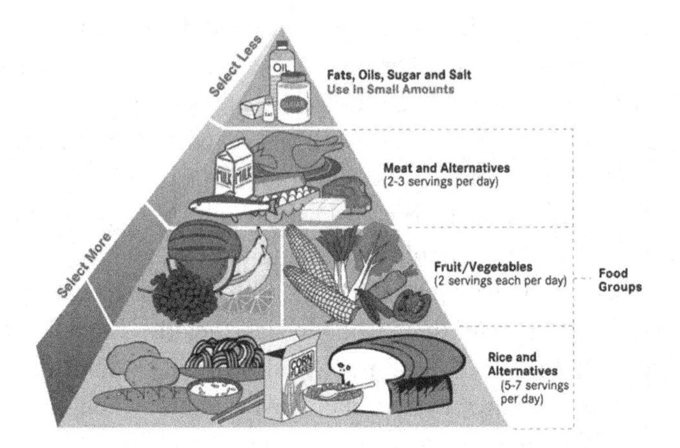

Figure 1. Recommended food servings for adults (http://www.hpb.gov.sg/)

2.3 *Activity three: Measuring and estimating serving sizes*

This activity is a practical one that involves measuring, weighing, estimating and recording. Students will be discussing the importance of healthy food and the fact that people eat a variety of foods to keep them healthy. They might be considering which foods are always good to eat and some that are to be eaten occasionally. The numbers of recommended servings for adults for the different categories are given on the pyramid, but not for young people (and these vary with age). Table 2 provides some examples of serving sizes from the Health Promotion Board of Singapore. In Activity three, students will consider the proportions of food that they need for their age group.

Teachers will decide how to implement this activity in the classroom. The Measurement section of the Mathematics syllabus gives some guidelines of levels of the activity suitable at different ages. In Primary 2 the students are involved in the estimation and measurement of length in metres and centimetres, mass in kilograms and grams, and volume of liquid in litres (in Primary 3 they are introduced to millilitres).

The students will need different food types, measuring jugs and kitchen scales to produce representatives of these serving sizes. A good practice is to ask students to estimate before measuring, to improve both estimation and measuring skills. Students should work in pairs or groups and discuss the process they are taking to produce the serving sizes. This activity should help students to become aware that serving sizes will never be exact; there will be errors in the measuring process and differences in the size of fruit and other products. They should be asking questions about the size of a cup or bowl and whether it really matters. For example a "palm size" can vary depending on the size of your hand, so students might ask why has this unit been used?

Table 2

Looking at serving sizes (http://www.hpb.gov.sg/)

Rice and alternatives	**Vegetables**
2 slices bread (60g) ½ rice bowl of rice (100g) ½ bowl beehoon(100g) 4 plain biscuits (40g) 1 thosai (60g) 2 small chapattis (60g) 1 large potato (180g)	¾ mug cooked leafy veg (100g) ¾ mug cooked non-leafy veg (100g) ¼ round plate 10" cooked veg 150g raw leafy vegetables 100g raw non-leafy vegetables
Fruit	**Meat and alternatives**
1 small apple, orange, pear, mango 1 wedge pineapple, papaya, watermelon 10 grapes or logas 1 medium banana 1 glass fruit juice (250ml)	1 palm size piece fish lean meat or skinless poultry 2 small blocks beancurd (170g) ¾ cup cooked pulses (120g) 5 medium prawns(90g) 3 eggs (150g) 2 glasses milk (500ml) 2 slices cheese (40g)

This activity can extend to estimating the weights of other objects to help students improve their estimation and measuring skills.

2.4 *Activity four: Developing a healthy menu*

The students now have information about what they should be eating from the Healthy Diet Pyramid and some information about serving sizes. The requirements for different age groups recommended by the Health Promotion Board of Singapore are in Table 3. Students start with serving sizes for adults in a selection of healthy foods that will be part of their menu. They may work in pairs even though they may develop a menu for themselves; as discussion can help to clarify the mathematics involved.

Table 3
Serves per day for Singapore youth (http://www.hpb.gov.sg/)

Food	7-12mths	1-2yrs	3-6yr	7-12yr	13-18yr
Rice/alternatives	1-2	2-3	3-4	5-6	6-7
Fruit	½	½-1	1	2	2
Veg	½	½	1	1	2
Meat/alternatives	½	½	1	2	2
Milk	750ml	750ml	500ml	250-	250-

In this activity students decide on their chosen food for a day's menu (or longer for older students) and then calculate the serving sizes according to their age group. They should record the serving sizes for each food item. They can start to use fractions of food servings to make up the required number of serves per category. For example, they may choose to include half-serves of two different vegetables to make one serve. This can be easier or more complex depending on the scope that students are given. For younger students they will need more scaffolding and some clear examples of what they need to do. The teacher will be able to decide whether they could be given a set of foods to choose from or given more free choice, depending on the background of the students. This activity can be extended to looking for recipes and so involve fractions and more complex calculations for older students from Primary 5.

2.5 *Activity five: Analysing the nutrition information panels*

In most countries packaged food has a nutritional information panel (NIP) that provides information about the energy in kilojoules (kJ) or calories (to convert kJ to calories divide by 4.2, or 4 for a quick estimate), fats, proteins, carbohydrates, sugars, sodium (to convert sodium to salt multiply by 2.5) and sometimes fibre and gluten. In addition, Singapore packets may display a Healthier Choice logo from the Singapore Heart Association. Australia has a Heart Foundation tick and the United Kingdom uses a traffic light system to help people make healthier choices. However, on some packets, or in advertisements, claims are made about the food that can often be quite misleading, and students can learn to look out for these. Most often they are to do with claims about "fat free" content that ignore, or attempt to "hide", the high levels of sugar put in to compensate for low fat content.

The following activities are suitable for Primary 4 to 6 students who are learning about units, proportions, percentages and reading tables. The activities firstly require the students to bring in from home clean food packets with nutritional information printed on them. These might be from food products such as noodles, vegetables, prepared meals, biscuits and snacks. The packets are cut up to make them usable, while retaining the essential information, and put into a zip lock plastic bags, or boxes to make 'Food Kits'. Each Food Kit should contain labels from a range of different foods.

These activities help students to read and interpret tables of information from which they can then make decisions. Some of the tables are quite dense with both nutritional information and percentage of daily allowance. Some labels on breakfast cereals even include information about consumption with and without milk, requiring more complexity in reading. Reading tables and graphs is an essential part of understanding our everyday lives. A *Five Step Framework*, with examples, to help students to develop the skills of interpreting tables and graphs is provided in Kemp (2010).

The following activities are part of Activity five and they provide some contexts for students to examine the Nutrition Information Panels in some detail. These panels have a variety of different formats and contain different information and students can find it helpful to actually have a purpose in looking at the panels. If students have already done Activity three where they were engaged in measuring weights and capacities they will already have some experience of the units such as grams and millilitres that can be seen in the Nutrition Information Panels. In Primary 5 students are looking at units in more sophisticated ways including conversion of units to bigger and smaller units and can engage in converting between the units where appropriate.

Traffic lights

The system of traffic lights, as used in the United Kingdom is a useful one enabling people to choose quickly in the supermarket. This is a good introduction to looking at labels. Students need to locate the "per 100g" column in the label which is different from the "per serving" column. When looking at the numbers students are deciding on whether the packet number fits into the ranges in Table 4. In so doing they are developing a mathematical concept and skill of deciding. This is of course more difficult when decimals are involved and students may need more guidance there.

Table 4

Traffic light colours for fats, sugar and sodium

Total fat (g/100g)	Green ≤3 O	Amber 3.1-19.9 O	Red ≥20 O
Saturated fat (g/100g)	Green ≤1.5 O	Amber 1.6-14.9 O	Red ≥5 O
Sugars (g/100g)	Green ≤5 O	Amber 5.1-14.9 O	Red ≥15 O
Sodium (mg/100g)	Green ≤120 O	Amber 121-599 O	Red ≥600 O

Source: Choice: The people's watchdog (October 2011, p.39)

Students can work in pairs with their Food Kits to allocate traffic light colours to the items. They should be encouraged to note anything they find surprising about the food and discussion following the activity should highlight these. Other points to highlight will be the differences between the brands and the salt and salt reduced versions of the same food. It is useful to have some discussion of how the recording will take place.

What can I eat if I have Diabetes type 2?

Research at the National University of Singapore by Seng and Lin (2011) has found that "Singapore has one of the highest incidences of diabetes in the developed world and …we hope to discover effective prevention and early intervention strategies, which may be in the form of simple lifestyle and nutritional interventions or perhaps even prophylactic drugs".

The occurrence of Diabetes type 2 is increasing worldwide and this is impacting on many families. Students, using a different Food Kit for this activity, can investigate the kinds of packaged foods that someone with Diabetes type 2 can eat. A general guideline that can be used with primary students is:

Not more than 5 grams of fat per 100 grams of food
AND not more than 10 grams of sugar per 100 grams of food

They can complete a worksheet or make up tables of their own to complete this activity. This activity involves making comparisons and judgements about two aspects of the nutritional information panels. Again, they will be surprised by the contents of some of the foods.

Percentage of daily requirements

On some packets this is included in the Nutritional Information Panels and on others it is given on the packet, outside of the panel. For students studying percentages and learning to construct tables and graphs it can be an interesting and challenging activity to look at these.

From a popular breakfast cereal on the top of the packet we can see that for a 40 gram serving the percentages of daily requirements are as follows in Table 5:

Table 5

Percentage of daily requirements per 40 gram serving for adults

Energy	**Fat**	**Sat Fat**	**Sugar**	**Sodium**	**Fibre**
640 kJ	2.5 g	0.5 g	9.1 g	100 mg	3.2 g
7%	4%	2%	10%	4%	11%

Although not stated on the packet these daily allowances are presumably for adults so some consideration would need to be given to the age of the students. There is worthwhile discussion about these percentages and what they mean, as students will see this on the packets.

Making direct comparisons between Nutritional Information Panels

As mentioned earlier there are differences between products within the same brand as well as across brands and students can make some comparisons across the packets of similar foods in the Food Kits. An example of the sorts of comparisons that can be made is provided in Table 6 which is adapted from an example in Kemp (2007).

Table 6

Making direct comparisons between Nutrition Information Panels

ORIGINAL CRISPBREAD Premium

NUTRITIONAL INFORMATION
Servings Per Package: 8
Serving Size: 30g (Approx. 5 biscuits)

	Average Quantity Per Serving	Average Quantity Per 100g
Energy	575kJ	1915kJ
Protein	2.5g	8.3g
Fat -Total	4.8g	16.1g
-Saturated Fat	2.3g	7.7g
Carbohydrate -Total	20.4g	67.9g
-Sugar	0.7g	2.4g
Sodium	249mg	829mg

98% FAT FREE CRISPBREAD Premium

NUTRITIONAL INFORMATION
Servings Per Package: 7
Serving Size: 30g (Approx. 5 biscuits)

	Average Quantity Per Serving	Average Quantity Per 100g
Energy	484kJ	1613kJ
Protein	3.0g	9.9g
Fat -Total	Less than 1g	1.6g
-Saturated Fat	Less than 1g	Less than 1g
Carbohydrate -Total	23.9g	79.8g
-Sugar	Less than 1g	Less than 1g
Sodium	152mg	505mg

98% FAT FREE HIGH FIBRE Premium CRISPBREAD

NUTRITIONAL INFORMATION
Servings Per Package: 7
Serving Size: 30g (Approx. 5 biscuits)

	Average Quantity Per Serving	Average Quantity Per 100g
Energy	467kJ	1556kJ
Protein	3.0g	9.9g
Fat -Total	Less than 1g	1.8g
-Saturated Fat	Less than 1g	Less than 1g
Carbohydrate -Total	21.9g	72.8g
-Sugar	1.1g	3.5g
Dietary Fibre	3.1g	10.2g
Sodium	142mg	473mg

A shopper looking to buy a packet of crisp bread suitable for a person with Diabetes type 2 observes that for this brand there are three versions: Original, 98% Fat Free and 98% Fat Free High Fibre. Which one (if any) would you choose and why?

Premium Original The total fat content of 16.1g is above the 5g per 100g requirement and so is too high, but the sugar at 2.4g per 100 g falls into the guidelines of less than 10g per 100g, sodium 829mg per 100g. The product is not suitable.

Premium 98% Fat Free (LOW IN SUGAR) The total fat content of 1.6g per 100g confirms the claim on the packet of being 98% fat free. The sugar content is less than 1g per 100g and the product is acceptable for a person with Diabetes type 2 (sodium 505mg per 100g).

Premium 98% Fat Free High Fibre (HIGH FIBRE) The total fat content of 1.8g per 100g confirms the claim on the packet of being 98% fat free. The sugar content is 3.5g per 100g and thus the product is acceptable for a person with Diabetes type 2. However, it may be noted that this is an increase of 1.1g of sugar per 100g on the *Original* version (sodium 473mg per 100g).

None of the packets claim to be 'low-salt' and so do not need to conform to the requirement of not more than 100mg of sodium per 100g; nevertheless it can be seen that the sodium content decreases from *Original* (829mg) to *98% Fat Free* (505mg) and to *98% Fat Free High Fibre* (473mg per 100g).

2.6 *Activity six: Using technology for making comparisons*

There is quite a lot of information on the panels and making comparisons and decisions can be quite complex. The use of technology to make comparisons allows students to not only make more detailed comparisons but also develop their skills in using technology. This would follow the rationale in the Mathematics Primary Syllabus (MOE, 2006a, p. 2) which states “Emphasis is also given to reasoning, applications, and use of technology. Advances in technology have changed the way we teach and learn mathematics. The computer and hand-held calculator, for example, offer great potential to enhance the teaching and learning of mathematics”.

Using spreadsheets for comparing packaged food products
Students can enter the numerical data into a spreadsheet on the computer and sort by the chosen category so that several categories can be compared. Some examples of savoury biscuits are included in the spreadsheet in Table 7. These have been sorted by fat content. The table makes clear that there is a large range for this variable and that biscuits that are high in fat may well be high in other components too. The students can sort by whichever component that they choose and they will need to make decisions about what they want to find out.

Table 7

Comparing savoury biscuits, sorted by total fat content in Excel

Brand	Product	Energy (Kj)	Protein (g)	Fat (tot g)	Fat (sat g)	Carbohydrates (g)	Sugars (g)	Sodium (mg)
W/Watchers	Crisbread	1340	9.1	1.2	0.2	74.5	1.9	316.0
Arnott's	Cruskits Rye	1390	9.1	1.5	0.3	63.6	1.6	640.0
Nabisco	Premium Crisbread	1560	10.8	1.5	0.6	76.0	4.3	830.0
Kraft	98% Fat Free	1609	9.9	1.5	0.5	79.9	0.5	506.0
Nabisco	Premium 98% Fat Free	1613	9.9	1.6	0.5	79.8	0.5	505.0
Kavli	Crispbread	1421	9.8	1.7	0.3	70.2	6.3	350.0
Nabisco	98% Fat Free High Fibre	1556	9.6	1.8	0.5	72.8	3.5	473.0
Kraft	Premium High Fibre	1529	10.0	1.9	0.5	70.6	2.6	668.0
Ryvita	Original Rye	1447	8.6	2.1	0.4	64.2	1.3	400.0
Trident	Rice Crakers	1660	8.5	2.2	1.0	85.3	5.0	408.0
Paradise	Lites	1586	10.9	2.9	0.6	73.0	3.2	360.0
Real Foods	Corn Thins	1593	9.5	3.0	0.0	77.6	0.5	201.0
Trident	Rice Crakers - Cheese	1720	7.0	3.1	1.5	87.5	0.5	120.0
Fantastic	Cheese Rice Crakers	1706	7.0	4.3	2.1	84.0	3.8	426.0
Trident	Rice Crakers S/Cream & Ch	1750	8.4	4.3	2.2	85.4	0.1	260.0
Trident	Rice Crakers - French Onion	1760	8.7	4.3	2.1	85.2	0.1	200.0
Water Thins	Sesame	1650	10.8	4.4	1.8	76.6	8.3	1220.0
Ryvita	Country Grains	1401	10.0	5.2	0.9	61.1	3.8	400.0

This task is certainly within the capabilities of students at upper primary school and this kind of activity gives students the responsibility to own their own research with their chosen food products. This can lead into oral presentations and discussions of the food claims on the packets.

2.7 *Activity seven: Body Mass Index calculations and analysis*

This activity includes students performing calculations, looking at ranges and percentiles on a graph and coming to a conclusion about the category a student fits: underweight, acceptable weight, overweight and obese. This can be done using hypothetical cases as examples and then the students in the class can calculate theirs for themselves. Some sensitivity would be needed in the latter case, especially if there are students in the class who fit into the extremes.

The Body Mass Index (BMI = Weight in kg $\div$ (height in m)^2) is often used as a measure of how your weight fits into a range and how close you are to your ideal weight. The recommended values vary between different countries and are not necessarily the best predictor of overall health. Some medical sources argue that for adults the waist measurement is more predictive of ill health as the fat deposited around the waist affect the organs in that area. However, it is probably one contributing factor as Diabetes UK (2011) proposes that reliable predictors of diabetes are one or more of: (i) a first degree family history of diabetes (ii) overweight or obese or morbidly obese with a BMI of 25 kg/m^2 and above, with a sedentary lifestyle (iii) a waist measurement > 94cm for men, and > 80cm for women. The calculation of BMI is the same for adults and children.

Naturally people vary in their body structure and so a range of weights is given. The Heart Foundation of Singapore (2011) tells us that "for a person whose body weight exceeds his recommended weight range by 20% or more, the risk of a heart attack is three times greater than for a person whose weight is in the healthy weight range. This is because being overweight or obese increases one's chances of developing other contributory factors for cardiovascular disease". According to the Singapore Heart Foundation, the BMI of Singaporean adults should ideally be between 18.5 and 22.5. A BMI < 18.5 being underweight, 23.0-29.9 overweight and >30 obese. The BMI range for some other countries vary a bit from these.

Although, the BMI for children is calculated in the same way as for adults, the ideal weight ranges vary according to age. A Health Promotion Board online Fact sheet BMI-for-age charts for children (2011) includes the information that teachers and students need to use to work out the categories of severely underweight, underweight, acceptable weight, overweight or seriously overweight applicable to students of different weights and for different ages. It includes the ranges for boys and girls and graphs of the percentiles as in Figure 2.

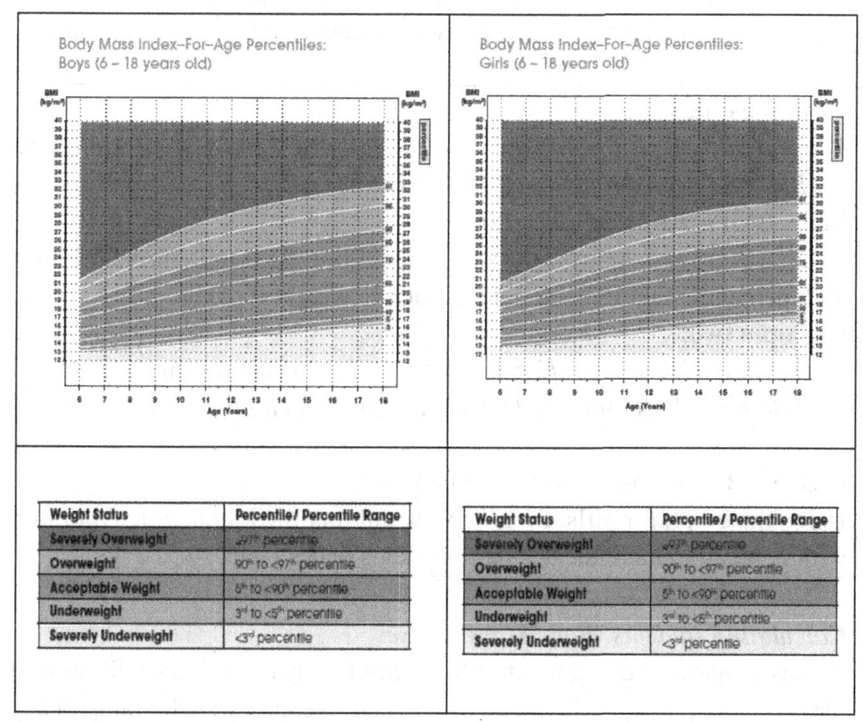

Weight Status	Percentile/ Percentile Range
Severely Overweight	≥97th percentile
Overweight	90th to <97th percentile
Acceptable Weight	5th to <90th percentile
Underweight	3rd to <5th percentile
Severely Underweight	<3rd percentile

Weight Status	Percentile/ Percentile Range
Severely Overweight	≥97th percentile
Overweight	90th to <97th percentile
Acceptable Weight	5th to <90th percentile
Underweight	3rd to <5th percentile
Severely Underweight	<3rd percentile

Figure 2. Analysing the Body Mass Index for boys and girls

The following examples use these steps for girls or boys of different ages and weights.

Step 1: Calculate BMI= Weight in kg ÷ (height in m × height in m)

Step 2: Select the appropriate BMI-for-age chart according to gender. Find the BMI on the vertical axis. Draw a horizontal line across from that point.

Step 3: Find the age on the horizontal axis. Draw a vertical line upwards from that point.

Step 4: Find the point where the vertical and horizontal lines meet. This indicates the BMI percentile or BMI percentile range where the BMI falls within.

Step 5: Match the BMI percentile reading or BMI percentile range to the corresponding weight status table to determine the weight status.

Example one:

Ben is a boy who is 9 years old, is 1.2 m tall and weighs 30 kg. Calculate his BMI and decide if his weight is acceptable.

Ben's BMI = $30 \div (1.2 \times 1.2) = 20.83$

Using the graph for boys, for a 9 year-old this BMI falls in the 5th to < 90th percentile that indicates an acceptable weight for Ben.

Example two:

Nicole is a girl who is 12 years old, is 1.3 m tall and weighs 45 kg. Calculate her BMI and decide if her weight acceptable.

Nicole's BMI = $45 \div (1.3 \times 1.3) = 26.63$

Using the graph for girls, for a 12 year-old this BMI falls in the 90th to < 97th percentile that indicates that Nicole is overweight.

Calculating students' Body Mass Indexes

This is a practical component to this activity where students will need to measure their heights in metres and their weights in kilograms. They then need to calculate their BMI using this information. Once they have the BMI and their age they can use the graphs in Figure 2 to find their category. The reading of the graph and understanding percentiles is a complex task and therefore suitable for the older students.

3 Conclusion

This chapter has suggested ways in which teachers might use everyday context in their classrooms in the middle to upper primary school. It drew on the context of health for the examples because this is an important part of everyday life; other contexts could be chosen relevant to the students' interests. The message for teachers is that they can motivate students to learn and use mathematics to engage with everyday contexts so that they can understand them better and make their own lifestyle decisions with health in mind.

References

Clemons, R. (2011, October). The whole truth. *Choice: The people's watchdog* (pp. 38-39).

Diabetes UK. (2011). Retrieved November 10, 2011, from http://www.diabetes.org.uk.

Dindyal, J. (2009). *Applications and modelling for the primary mathematics classroom.* Singapore: Prentice Hall.

Gal, I. (2004). Statistical literacy. In D. Ben-Zvi & J. Garfield (Eds.), *The challenge of developing statistical literacy, reasoning and thinking* (pp. 47-78). Dortrecht, The Netherlands: Kluwer Academic Press.

Health Promotion Board. (2011). HPB Online. Retrieved November 12, 2011, from http://www.hpb.gov.sg/

Health Promotion Board. (2011). *A fact sheet BMI-for-age charts for children.* Retrieved November 18, 2011, from http://www.hpb.gov.sg/uploadedFiles/HPB_Online/News_and_Events/News/2010/fact%20sheet%20bmi.pdf

Kaur, B. & Yeap, B. H. (2009). *Pathways to reasoning and communication in the primary school mathematics classroom: A resource for teachers by teachers.* Singapore: National Institute of Education.

Kemp, M. (2007). Mathematics and food: Essentials for life. In K. Milton, H. Reeves, & T. Spencer (Eds.), *Mathematics: Essential for learning, essential for life – Proceedings of the 21st Biennial Conference of the Australian Association of*

Mathematics Teachers Inc. (pp. 322-330). Australia: Australian Association of Mathematics Teachers Inc.

Kemp, M. (2010). Developing pupil's analysis and interpretation of graphs and tables using a *Five Step Framework.* In B. Kaur & J. Dindyal (Eds.), *Mathematical applications and modelling: Yearbook 2010* (pp.199-218). Singapore: World Scientific Publishing Company.

Ministry of Education. (2006a). *Mathematics syllabus – Primary.* Singapore: Author.

Ministry of Education. (2006b). *Mathematics syllabus – Secondary.* Singapore: Author.

Ministry of Education. (2007). *Health education syllabus for primary level.* Singapore: Author.

Seng, C., & Lin, L. (2011). *Fighting diabetes and other metabolic diseases in Singapore.* Retrieved November 13, 2011, from http://www.nus.edu.sg/research/rg156.php

Singapore Heart Foundation. (2011). *Singapore heart foundation.* Retrieved November 12, 2011, from http://www.myheart.org.sg

Steen, L. A. (Ed.). (1997). *Why numbers count: Quantitative literacy for tomorrow's America.* New York, NY: The College Board.

Watson, J. (1995). Statistical literacy: a link between mathematics and society. In A. Richards (Ed.), *Forging links and integrating resources* (pp. 12-28). Darwin: The Australian Association of Mathematics Teachers.

Zevenbergen, R., Doyle, D., & Wright, J. (2004). *Teaching mathematics in primary schools.* NSW, Australia: Allen & Unwin.

Chapter 15

Mathematics, Astronomy and Culture: Helping Students See Connections

Helmer ASLAKSEN

In this chapter some interesting mathematics content from undergraduate mathematics courses that shows the connection and relevance of mathematics to daily life is presented. The philosophy behind these mathematics courses is also discussed. Hopefully, teachers will be inspired to develop exciting material for their mathematics classroom teaching.

1 General Education Modules at the National University of Singapore

It has always been my dream to inspire undergraduate students with mathematical content knowledge that is fun, and that shows the connection between mathematics and the real world. In this chapter I will present some of the content and philosophy of the two courses: 1) Heavenly Mathematics and Cultural Astronomy; and 2) Mathematics in Art and Architecture. I first taught these courses at the National University of Singapore (NUS) in 2000. These two courses allow me to fulfill my dream to excite undergraduate students with mathematical knowledge they will treasure for the rest of their lives and that they will pass on to their children, and knowledge that will make them see the beauty and significance of mathematics, and make them look at the world around them with different eyes. The objectives of both courses are:

1) facilitate the students appreciate the world around them and start looking at their surroundings with different eyes, so that they will begin to notice and question things they used to take for granted;
2) show them connections between mathematics and culture so that they will switch from thinking about mathematical content knowledge in terms of school subjects to the knowledge begin part of mankind's struggle to understand the world; and
3) demonstrate the relevance and importance of mathematics by showing how it solves problem of general interest, so that they could appreciate the beauty and centrality of mathematics.

Many mathematics educators have worked on developing connections between mathematics and other subjects. This was the topic of the 1995 Yearbook of the NCTM (House & Coxford, 1995) and a Topic Study Group at the 10th International Congress on Mathematics Education (Anaya & Michelsen, 2004). My goal in this chapter is to show the connection between mathematics and astronomy. Astronomy is a topic that appeals to many students, lends itself easily to hands-on activities, and is often not covered well in the school curriculum. Furthermore, astronomy is "local", in the sense that students in the tropics will see different things from students in the temperate zone.

2 Selected Topics from My Astronomy Course

The course Heavenly Mathematics and Cultural Astronomy begins with basic astronomy, focusing on how the Sun, the Moon and the stars look like from different parts of the world (Aslaksen, 2011). While many astronomy books take a "high latitude centric" point of view, this course attempts to be "hemispherically correct" and focus on the tropical point of view. The course then applies this to applications of astronomy with a cultural flavor, like calendars, navigation and sundials.

2.1 *Calendars in Singapore*

In this section, the use of mathematics to determine the public holidays in Singapore is discussed. Public holidays are determined using the Gregorian, Chinese, Islamic and Indian calendars. The Gregorian calendar is arithmetical and trivial, while the three others are based on actual astronomical events and are therefore complex (Dershowitz & Reingold, 2007). There are 11 public holidays in Singapore. Three of them are secular, January 1, Labor Day and National Day, while the remaining eight consist of two Chinese (Chinese New Year and the second day of Chinese New Year), two Muslim (Eid ul-Fitr, called Hari Raya Puasa in Malay, and Eid ul-Adha, called Hari Raya Haji in Malay), two Indian (Vesak Day or Buddha's birthday and Deepavali) and two Christian (Good Friday and Christmas Day).

The Earth revolves around the Sun and this determines the tropical year, which is 365.2422 days. The synodic month is the time from one new Moon to the next. The average length is 29.53 days. Since

$$365 - 12 \times 29.5 = 11,$$

a lunar year consisting of 12 lunar months is about 11 days short of a solar year. After 2 years, the error is about 22 days, and after 3 years about 33 days. In order to prevent Chinese New Year from moving backwards through the calendar (like the Muslim holidays do), a leap month is inserted about every third year. That is why the Chinese calendar is not a lunar calendar, but a lunisolar calendar. Any month can be followed by a leap month. For more details about the Chinese calendar, see Dershowitz and Reingold (2007), Aslaksen (1999) and Aslaksen (2002).

The Chinese seasonal marker lìchūn (立春) marks the beginning of spring and falls on February 4 (or 5). It is halfway between the spring equinox and the winter solstice. Chinese New Year is the new Moon closest to lichun. It falls between January 21 and February 21. Normally, Chinese New Year moves 11 days (or 10 or 12) earlier once or twice, but if an 11-day step would take it before January 21, we have to add a leap month, so it jumps 30 – 11 = 19 (or 18 or 20) days later. This is illustrated in Table 1.

Table 1

The Movement of Chinese New Year

2005	Feb 9	
2006	Jan 29	-11
2007	Feb 18	+20
2008	Feb 7	-11
2009	Jan 26	-12
2010	Feb 14	+19
2011	Feb 3	-11
2012	Jan 23	-11
2013	Feb 10	+18
2014	Jan 31	-10

The Chinese months start on the day of the new Moon, but at the time of the new Moon, we cannot see the Moon. The Muslim calendar starts with the first visibility of the lunar crescent, which will occur after sunset a day or two after the new Moon. Another complication is that the Chinese day runs from midnight to midnight, while the Muslim day starts at sunset. So if for example the crescent is seen after sunset on the second day of the Chinese month, the first day of the Muslim month will be the third day of the Chinese month.

Deepavali is a national public holiday in India, Sri Lanka, Nepal, Mauritius, Fiji, Malaysia and Singapore. The rules are somewhat different in the northern and southern parts of India, so the holiday in Sri Lanka, Malaysia and Singapore will fall one day earlier than the holiday in India and Mauritius about half the time. In Nepal both days are public holidays. For more details, see Aslaksen and Beltrami (2008).

In Singapore, the Hindu Endowments Board only makes an estimate of the date of Deepavali, and waits for calendars from India to appear before the date of the public holiday is confirmed by the Ministry of Manpower. Unfortunately, those estimates may not be correct, and the date for the holiday has to be revised after the start of the year. In Singapore, the date of Deepavali in 2009 was estimated to be 15 November 2009 on 21 February 2008. However, exact calculation showed that it should be 17 October 2009. I contacted both the Ministry of Manpower and the Hindu Endowment board in February 2008, and almost one year later, on 10 February 2009, they finally announced the change.

2.2 *The Singapore flag*

You have seen the Singapore flag your whole life, but have you noticed that it contradicts astronomical truths? The Moon is our nearest neighbour, so the stars inside the Moon will not be visible! A waxing crescent in Singapore will not look like the left crescent on the front of the flag or the right crescent on the back of the flag. In fact, it will look like the bottom crescent on the Singapore coat of arms! However, one good thing about the Singapore flag is that the horns of the crescent are diametrically opposite, unlike many other national flags that have horns that almost close up. Occasionally I see fake Singapore flags, like the one in Figure 1, where the stars form an upside down pentagon!

Figure 1. A "fake" Singapore flag

2.3 *Vertical sundials in the tropics*

The shadow of a stick placed vertically in the ground in the North Temperate Zone will appear to move in clockwise direction. The shadow of the stick in the South Temperate Zone will move in a counter-clockwise direction. But what about the same stick at the tropics? In the following explanation the source of Figures 2, 3, 4 and 5 is Sun (2011).

It can be shown that at latitude $x°$ north, the Sun will cross the horizon at an angle equal to $(90 - x)$ ° as shown in Figure 2. Suppose we are at latitude 10° north at the time of the June solstice. The Sun will then rise north of east, and cross the horizon at an angle of 80°, moving towards the south. The position of the Sun can be described by altitude and azimuth. Altitude is the angle above the horizon, while azimuth is the angle along the horizon, starting from the north point (see Figure 3). In our example, the azimuth will increase after sunset since the path of the Sun is tilting towards the south. However, since we are close to the equator, the path of the Sun is almost vertical, and it will cross the meridian $(23.5 - 10)$ ° = 13.5° north of zenith. Hence, before noon the azimuth will decrease. So in the morning, the azimuth of the Sun will increase, while before noon, the azimuth will decrease. It follows that the sundial changes direction twice in the course of the day! In Figures 4 and 5 these turning points are labeled T1 and T2.

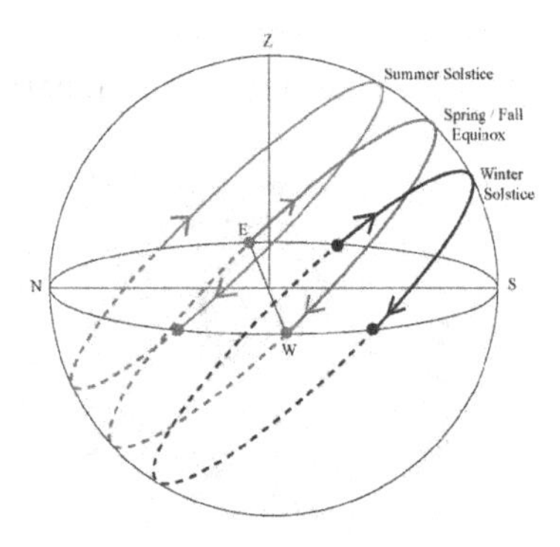

Figure 2. The path of the sun

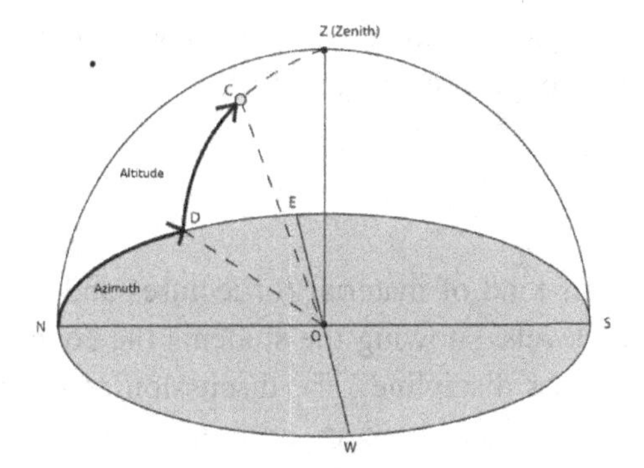

Figure 3. Altitude and azimuth

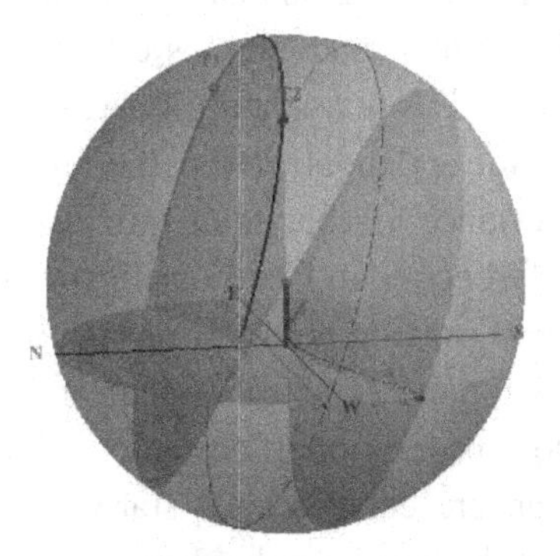

Figure 4. The path of the sun in the tropics

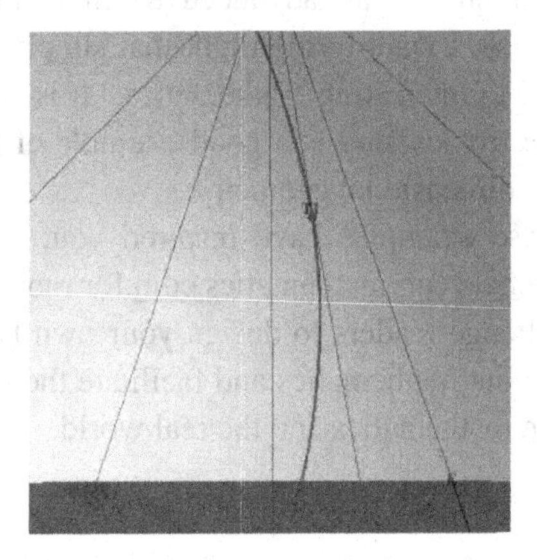

Figure 5. The sun turns in the tropics

Thus, in the tropics, making the sundial go backwards is normal. However, very few people are aware of this.

3 Conclusion

I have also used this kind of material for lectures and shorter courses at schools at various levels, showing the students the connections between mathematics and other discipline. The discussion of the Singapore flag stresses the importance of looking at familiar objects with critical, scientific eyes. Even primary school students are intrigued by this.

Students usually associate calendars with traditional culture, and enjoy seeing that science and mathematics can be applied to something they have earlier perceived as non-scientific. At first, some students may only be interested in the calendars from their own, race, culture and religion, but my experience is that they quickly become intrigued by seeing both the differences and similarities between the different calendars. This is a great starting point for project work. This can involve historical and cultural topics, or astronomical and mathematical topics. The students can do observational activities, like studying the first visibility of the lunar crescent, which starts the Islamic month, or computer simulation of the various calendars.

The sundial example is an advanced example meant for a more sophisticated audience. I want to highlight that simply putting a stick in the ground can lead to interesting mathematics. It is something that can only be done in the tropics, and is a good example of the difficulty and importance of three-dimensional geometry.

I hope that these examples have inspired you. Interested readers could refer to www.heavenlymathematics.com for more ideas and hands-on activities. I challenge readers to design your own hands-on activities to excite students about mathematics and facilitate them to appreciate the connection between mathematics and the real world.

References

Anaya, M. & Michelsen, C. (Eds.) (2004). *Relations between mathematics and other subjects of science or art, Proceedings of Topic Study Group 21 at ICME-10, The 10th International Congress on Mathematics Education, Copenhagen, Denmark, 2004*. Department of Mathematics and Computer Science, University of Southern Denmark.

Aslaksen, H. (1999). *The mathematics of the Chinese calendar.* Retrieved October 10, 2011, from http://www.math.nus.edu.sg/aslaksen/calendar/cal.pdf

Aslaksen, H. (2002). When is Chinese New Year?, *Griffith Observer 66*(2), 1-17.

Aslaksen, H. (2011). Heavenly Mathematics: Observing the sun and the moon from different parts of the world. In S. Tatiana, H. F. David, & A. L. Gerald (Eds.), *Expeditions in mathematics* (pp. 233-248). Washington D.C.: Mathematical Association of America.

Aslaksen, H., & Beltrami, O. (2008), *When is Deepavali (Diwali)?,* Retrieved October 10, 2011, from http://www.math.nus.edu.sg/aslaksen/calendar/deepavali.html.

Dershowitz, N. & Reingold, E. M. (2007). *Calendrical calculations.* Cambridge University Press.

House, P. A. & Coxford, A. F. (Eds.) (1995). *Connecting mathematics across the curriculum, 1995 yearbook.* Reston, VA: National Council of Teachers of Mathematics (NCTM).

Sun, Y. (2011). *Using 3D graphics to explain why a sundial can go backwards in the tropics.* Unpublished Honours Thesis, National University of Singapore, Singapore.

Contributing Authors

Helmer ASLAKSEN was born in Oslo, Norway, and did his undergraduate at the University of Oslo. After receiving his Ph.D. at the University of California, Berkeley, he joined the Department of Mathematics at the National University of Singapore in 1989. His interests include geometry, Lie theory, and the relationship between mathematics and astronomy and art. He has been academic advisor for the exhibition "Art Figures: Mathematics in Art" at the Singapore Art Museum, "The Dating Game: Calendars and Time in Asia" at the Asian Civilisation Museum and the TV series "Ancient Chinese Inventions" on the Discovery Channel. He was a judge for "National Science Challenge", a TV science quiz for secondary school students and "Brand's Asia Pacific Sudoku Challenge". He was on the organizing committee of a topic study group at the International Congress on Mathematical Education in 2004. He has been invited to be a plenary speaker for the Mathematical Association of America. He has an extensive web site, including a page on The Mathematics of the Chinese Calendar at www.chinesecalendar.org. At the NUS he has introduced two General Education Modules, "Heavenly Mathematics & Cultural Astronomy" and "Mathematics in Art and Architecture". In 2004 he was awarded the University's Outstanding Educator Award.

Melvin CHAN is a research associate and PhD candidate in Education at the Centre of Research in Pedagogy and Practice, National Institute of Education, Nanyang Technological University, Singapore. His dissertation is focussed on the study of self, identity and agency among Singaporean students. His research interests include the development of student capacities, concepts of the self, as well as the use of advanced statistical methodology such as structural equation modelling to examine these issues. Currently, Melvin is involved in a large-scale representative study on pedagogy and assessment practices in Singapore. Previously, he was involved in the life pathways project, a large-scale longitudinal study that examined students' academic and non-academic outcomes.

Jaguthsing DINDYAL is an Associate Professor at the National Institute of Education, Nanyang Technological University in Singapore. He holds a PhD in Mathematics Education from Illinois State University, USA. He has prior experience of teaching at the secondary level and currently teaches mathematics education courses to both pre-service and in-service teachers. He has been an EXCO member of the Association of Mathematics Educators since 2004 and is a life member of the Educational Research Association of Singapore. His research interests include reasoning and proofs, problem solving, geometric and algebraic thinking, international studies and the mathematics curriculum. His contributions in these areas include a number of refereed conference papers, book chapters and journal articles.

David John HOGAN is a Professor and Principal Research Scientist at the National Institute of Education in Singapore. Previously he was Dean of Educational Research, NIE and Dean of the Centre for Research in Pedagogy and Practice, NIE. Prior to that he held a series of academic appointments at the University of Tasmania (1994-2004) and the University of Pennsylvania (1979-1993). Currently he is Principal Investigator of the Core 2 Research Program funded by the Office of Education Research, NIE. The research reported in this project is based on this research program.

In-Ok JANG is a part-time Professor at the Chun-cheon National University of Education. She served as an elementary mathematics teacher for 20 years. She got her PhD at Korea National University of Education in 2010.

Berinderjeet KAUR, PhD, is a Professor of Mathematics Education and Head of the Centre for International Comparative Studies (CICS) at the National Institute of Education in Singapore. Her primary research interests are in the area of classroom pedagogy of mathematics teachers and comparative studies in mathematics education. She has been involved in numerous international studies of Mathematics Education and is the Mathematics Consultant to TIMSS 2011. She is the principal investigator (Singapore) of the Learner's Perspective Study (LPS) helmed by Professor David Clarke of the University of Melbourne. As the President of the Association of Mathematics Educators (AME) from 2004-2010, she has also been actively involved in the professional development of mathematics teachers in Singapore and is the founding chairperson of the Mathematics Teachers Conferences that started in 2005. She is also the founding editor of the AME Yearbook series that started in 2009. On Singapore's 41st National Day in 2006, she was awarded the Public Administration Medal by the President of Singapore.

Marian KEMP has been a mathematics educator at Murdoch University for over 25 years, firstly in the School of Education and then the Student Learning Centre where she was the Head of the Centre from 2004. She is currently the Director of Student Life and Learning at Murdoch University. Marian has provided support for undergraduates in mathematics and statistics across the university including developing their ability to make appropriate use of scientific and graphics calculators for learning mathematics. She has published in this field and has presented papers and workshops to teachers at conferences nationally and internationally. Marian has developed programs for improving numeracy across the curriculum that have involved critical numeracy tasks, and more recently a series of online numeracy modules, for students enrolled in Murdoch's first year interdisciplinary Foundation Units. Her research has involved the development of student strategies for interpreting graphs and tables, including the use of a Five Step Framework and in this field she has published papers and presented at conferences for teachers throughout Australia and internationally. In 2007 Marian was awarded a Carrick Institute Award for University

Teaching for outstanding contributions to student learning in the development of critical numeracy in tertiary curricula.

Barry KISSANE has been a mathematics educator at Murdoch University in Perth since 1985, except for a period working and studying at the University of Chicago and a recent period as the Dean of the School of Education at Murdoch University. His present teaching responsibilities include mathematics for primary teacher education students and mathematics education for secondary teacher education students. His research interests in mathematics education include numeracy, curriculum development, the use of technology for teaching and learning mathematics and statistics, popular mathematics, teacher education and others. He was written several books and many papers related to the use of graphics calculators in school mathematics, and published papers on other topics, including the use of the Internet and the development of numeracy. Barry has served as President of the Mathematical Association of Western Australia (MAWA) and as President of the Australian Association of Mathematics Teachers (AAMT). He has been a member of editorial panels of various Australian journals for mathematics teachers for around 30 years, including several years as Editor of The Australian Mathematics Teacher. A regular contributor to conferences for mathematics teachers throughout Australasia, he is an Honorary Life member of both the AAMT and the MAWA.

Dennis KWEK is a senior research associate at the Centre for Research in Pedagogy and Practice, National Institute of Education, Singapore. He is currently involved in a large-scale classroom observation research project utilising multidimensional qualitative and quantitative approaches to characterise teaching and learning in Singapore. Prior to this, he has researched issues of deficit thinking in Singapore classrooms, served as co-principal investigator for a major intervention study on building teacher pedagogical repertoires, as well as a collaborator for a research study on whole school reform of curriculum, assessment and pedagogy in a school for gifted and talented students. Before returning to Singapore in 2004, he has worked as a lecturer in the United Kingdom, with qualifications and research interests in Artificial Intelligence and Critical Management Theory. A doctoral candidate working on the notion of 'weaving', a pedagogical theory of connected learning across different types and forms of knowledge, values and skills, Dennis is also interested in discourse analysis, temporal analysis of classroom interactions, whole-school change, the impact of high stakes testing, teacher professional development, educational philosophy, and alternative pedagogical ideas from the East.

Kyeong-Hwa LEE is a Professor of Mathematics Education at Seoul National University. Her research and teaching interests embrace mathematics education, and particularly creativity development, gifted education, probability and statistics education, analogical reasoning, gender issues, and teacher education. She has been in charge of several research projects; for example, WISE (Women into Science and Engineering Project), APDMT (Advance Planning for Development of Mathematics Textbook), and Improvement of Teaching and Learning Methods Course in Mathematics to Establishing Practical Perspective. She has published 80+ journal articles, books, book chapters, reviews and commentaries in mathematics education. Among her recent major works include International Perspectives on Gender in Mathematics Education (The Information Age Publishing, 2010) and The Elements of Giftedness in Mathematics (Sense Publisher, 2011).

LEONG Yew Hoong is an Assistant Professor at the Mathematics and Mathematics Education Academic Group, National Institute of Education, Singapore. His research interest in the classroom enactment of quality mathematics instruction has led him to related areas of study which include mathematical problem solving, mathematics teacher development, and most recently, making connections in the teaching of mathematics.

Hee-Chan LEW is a Professor of Mathematics Education at the Korea National University of Education and the President of Korea Society of Educational Studies in Mathematics. He served as a member of

International Committee and International Program Committee of PME. He directed various researches in mathematics education on computer use in classroom, teaching methods and evaluation funded by Korea Research Foundation. He is an author of current high school textbooks and more than 100 research articles. He has been a presenter at various international conferences including AERA, NCTM, NCTM Research Session, PME, ICME, ATCM, EARCOME, APEC Lesson Study.

Tom LOWRIE is a Professor of Mathematics Education at Charles Sturt University, Australia and is Director of the Research Institute for Professional Practice, Learning and Education (RIPPLE). As Director, Tom coordinates the research activity of more than 40 key researchers and 60 doctoral students, focussing on professional practice, learning and education across disciplines. With an established international research profile in the discipline areas of mathematics education, his recent research has investigated the influence digital technologies may have on disadvantaged students' (particularly Indigenous students, and students living in remote areas). More broadly, Tom's work focuses on the extent to which children use spatial reasoning and visual imagery to solve mathematics problems and the role and nature of graphics in mathematics assessment. He has attracted over $1.6 million over 11 years in nationally-competitive research funding for projects on which he has held the position of Chief Investigator. Tom has disseminated his research in approximately 115 highly reputable national and international books, periodicals and conference proceedings.

Alwyn Wai-Kit PANG is an Assessment Specialist with the Singapore Examinations and Assessment Board (SEAB). Prior to his appointment in SEAB, he was a teacher in Anglo-Chinese Junior College. He holds a MSc in mathematics from National University of Singapore and a MEd in mathematics education from Nanyang Technological University.

Ridzuan Abdul RAHIM is a Senior Teaching Fellow at the Centre for Research in Pedagogy and Practice, NIE. He has postgraduate degrees in Mathematics and Statistics from the University of Cambridge and the National University of Singapore. He taught Mathematics at the junior college and secondary school levels, developed mathematics curriculum for the gifted programme and was an Assistant Director in the Gifted Education Branch before his secondment to NIE. He is currently completing his PhD.

Denisse R. THOMPSON is Professor of Mathematics Education at the University of South Florida in the United States. She received her PhD in Education from the University of Chicago in 1992, conducting an evaluation of the 12th grade course (Precalculus and Discrete Mathematics) developed by the University of Chicago School Mathematics Project (UCSMP); she has been involved with UCSMP for over 25 years, as an author, editor, and most recently as Director of Evaluation for the Third Edition materials. Dr. Thompson taught at the middle school, high school, and community college levels and joined the faculty in her current position in 1990. Her major scholarly interests are curriculum development and research, literacy in mathematics, and the integration of culture into the teaching of mathematics. She has authored or co-authored 17 books, over 25 book chapters, and over 50 journal articles and presented at over 100 conferences, including the International Congress on Mathematics Education, annual and regional meetings of the National Council of Teachers of Mathematics, and the East Asia Regional Conference on Mathematics Education. In 2004, she was named the Kenneth Kidd Educator of the Year by the Florida Council of Teachers of Mathematics, and in 2010 was named the Mathematics Teacher Educator of the Year by the Florida Association of Mathematics Teacher Educators.

TOH Tin Lam is an Associate Professor with the Mathematics and Mathematics Education Academic Group, National Institute of Education, Nanyang Technological University, Singapore. He obtained his PhD in Mathematics (Henstock-stochastic integral) from the National University of Singapore. Dr Toh continues to do research in mathematics as well as in mathematics education. He has papers published in international scientific journals in both areas. Dr Toh has taught in junior college in

Singapore and was head of the mathematics department at the junior college level before he joined the National Institute of Education.

Phillip Alexander TOWNDROW is a Senior Research Scientist in the Centre for Research in Pedagogy and Practice, National Institute of Education, Nanyang Technological University, Singapore. Phillip began his career in education as an English language teacher. He has worked in a variety of cultural and social contexts including Spain, the United Arab Emirates and most recently, Singapore. His research and writing interests include, curriculum and teaching design, new literacy studies, and teacher professional learning. Phillip has published widely in educational journals and has co-authored and co-edited three books on the use of Information Technology in English language learning, and two others in task design, and motivation theory and practice.